Agent Technology for
Communication Infrastructures

Agent Technology for Communication Infrastructures

Edited by

Alex L. G. Hayzelden and **Rachel A. Bourne**
Queen Mary, University of London, UK

JOHN WILEY & SONS, LTD
Chichester • New York • Weinheim • Brisbane • Singapore • Toronto

Other Wiley Editorial Offices

John Wiley & Sons, Inc., 605 Third Avenue,
New York, NY 10158-0012, USA

Wiley-VCH Verlag GmbH
Pappelallee 3, D-69469 Weinheim, Germany

Jacaranda Wiley Ltd, 33 Park Road, Milton,
Queensland 4064, Australia

John Wiley & Sons (Canada) Ltd, 22 Worcester Road
Rexdale, Ontario, M9W 1L1, Canada

John Wiley & Sons (Asia) Pte Ltd, 2 Clementi Loop #02-01,
Jin Xing Distripark, Singapore 129809

Library of Congress Cataloging-in-Publication Data

Agent technology for communication infrastructures /edited by Alex L.G. Hayzelden
and Rachel Bourne.
      p. cm.
    Includes bibliographical references and index.
    ISBN 0 471 49815 7
        1.   Intelligent agents (Computer software) 2. Telecommunication systems. I.
    Hayzelden, Alex L. G., 1974- II. Bourne, Rachel

    QA76.76.158 A34 2000-10-31
    006.3 – dc21   00-047760

British Library Cataloguing in Publication Data

A catalogue record for this book is available from the British Library

ISBN 0 471 49815 7

Produced form Word files supplied by the authors
Printed and bound in Great Britain by Biddles Ltd., Guildford and Kings Lynn
This book is printed on acid-free paper responsibly manufactured from sustainable forestry,
in which at least two trees are planted for each one used for paper production.

Contents

Contributors

The editors would like to thank all those who were involved in aspects of contributing to the chapters for this book; they include:

NAME	AFFILIATION
P. Alimonti	Fondazione Ugo Bordoni, Italy.
H. Almiladi	Teltec DCU, Ireland.
J. Barria	Imperial College of Science, Technology and Medicine, UK.
G. Bayless	British Telecommunications plc, Reston, VA, USA.
J. Bigham	Queen Mary, University of London, UK.
E. L. Bodanese	Queen Mary, University of London, UK.
A. Bordetsky	Information Systems, Naval Postgraduate School, USA.
J. Borrell	Universitat Autònoma de Barcelona, Spain.
R. A. Bourne	Queen Mary, University of London, UK.
P. Buckle	Nortel Networks, plc. UK.
T. Curran	Teltec DCU, Ireland.
L. G. Cuthbert	Queen Mary, University of London, U.K.
B. Davis-Omene	British Telecommunications Plc., Reston, VA, USA.
R. Evans	Broadcom Éireann Research Ltd., Dublin, Ireland.
R. Fabregat	Universitat de Girona, Spain.
N. Fujino	NetMedia Research Center Fujitsu Laboratories Ltd, Japan.
C. Gerber	German Research Center for Artificial Intelligence (DFKI), Germany.
M. A. Gibney	Queen Mary, University of London, U.K.
J. M. Griffiths	Queen Mary, University of London, U.K.
M. Hansen	Technical University of Denmark.
D. Harle	University of Strathclyde, U.K.
A. Hayzelden	Queen Mary, University of London, U.K.
J. Hickie	Broadcom Éireann Research Ltd, Dublin, Ireland.
I. Iida	NetMedia Research Center Fujitsu Laboratories Ltd, Japan
N. R. Jennings	Queen Mary, University of London, U.K.
P. Jenson	Tele Danmark, Denmark.
R. Kerkdijk	KPN Research, The Netherlands.
D. Kerr	Broadcom Éireann Research Ltd., Dublin, Ireland.
M. Klusch	German Research Center for Artificial Intelligence, Germany.

O. Krone	Swisscom Corporate Technology, Switzerland and Berkeley University, USA.
S. Lloyd	Queen Mary, University of London, UK.
F. Lucidi	Fondazione Ugo Bordoni, Italy.
E. Lukschandl	Chalmers University of Technology, Sweden.
Z. Luo	Queen Mary, University of London, UK.
M. Mäkelä	University of Helsinki, Finland.
J. L. Marzo	Universitat de Girona, Spain.
S. Mathews	Broadcom Éireann Research Ltd., Dublin, Ireland.
Y. Matsuda	NetMedia Research Center Fujitsu Laboratories Ltd, Japan.
B. T. Messmer	Swisscom Corporate Technology, Switzerland.
P. Misikangas	University of Helsinki, Finland.
T. Nishigaya	NetMedia Research Center Fujitsu Laboratories Ltd, Japan.
J. Odubiyi	British Telecommunications plc., Reston, VA, USA.
D. O'Sullivan	Broadcom Éireann Research Ltd, Dublin, Ireland.
A. Patel	Imperial College of Science, Technology and Medicine, UK.
J. Pitt	Imperial College of Science, Technology and Medicine, UK.
S. J. Poslad	Imperial College of Science, Technology and Medicine, UK.
K. Raatikainen	University of Helsinki, Finland.
S. Robles	Universitat Autònoma de Barcelona, Spain.
D. Rossier-Ramuz	Swisscom Corporate Technology, Bern, Switzerland.
E. Ruberton	British Telecommunications Plc., Reston, VA, USA.
Y. Sakai	Tokyo Institute of Technology, Japan.
R. Scheurer	University of Fribourg, Switzerland.
J. Soldatos	National Technical University of Athens, Greece.
S. Sugawara	The University of Electro-Communications, Japan.
S. Trigila	Fondazione Ugo Bordoni, Italy.
E. Vayias	National Technical University of Athens, Greece.
P. Vilà	Universitat de Girona, Spain.
N. J. Vriend	Queen Mary, University of London, UK.
K. Yamaoka	Tokyo Institute of Technology, Japan.

Preface

Agent technology has been demonstrated to be a useful tool for developing aspects of communications infrastructures, from enabling the deployment of novel e-business application layer services to tackling the broad issues inherent in the underlying control system. The editors are pleased to announce a broad mix of technical contributions from Asia, Europe and the US, that cover a variety of issues related to the these areas.

Separate chapters in the book focus on the use of software agent technology to a range of applications, from traditional technical problems, such as resource allocation management, to advanced network control capabilities, such as the control of virtual home environments for third generation mobile-based services. Chapters are presented covering topics such as mobile agent technology, wireless application protocol, e-commerce, and various advanced network management approaches. The main objectives of this book are to illustrate the diversity of communications applications using agent technology and to give an understanding of how the agent metaphor is used in practice.

At the time of writing, there is a huge amount of interest in e-business and e-commerce systems. However, the support of next-generation communication services is increasing the complexity of the communications infrastructure. Agent technology has featured as a powerful means of satisfying the important commercial need for software systems to interoperate and to manage large heterogeneous service platforms. Communications infrastructure is now seen as a natural application domain for agent technology. The chapters in this book describe and define how the application of agent technology has started to make an impact on e-business and the delivery of services in communications systems.

Conceptually, the book is divided into two parts. Part I, chapters 1–10, provides an introduction to agent technology in the area of service delivery in communications systems. The topics covered range from how agent technology can improve the issues of developing and maintaining novel communications services to the impact and improvement in security through the increased degrees of flexibility that using agents can provide. In particular, the first chapter of Part I provides the unacquainted reader with an overview of agent technology and describes some of the tools and development techniques used to facilitate its widespread uptake.

Part II, chapters 11–22, demonstrates how agent technology has been used to tackle various aspects of communications infrastructure to support the control and provisioning of value-added services. For example, the introduction of an increasing number of services leads to a similar increase in the underlying communications infrastructure complexity – agents are seen as a possible solution for dealing with the associated issues.

Many of the chapters have been written by authors who collaborated on the European ACTS IMPACT project. More information can be found at: http://www.acts-impact.org/

Alex Hayzelden and **Rachel Bourne,** August, 2000

Acknowledgements

The editors would especially like to thank all of the invited contributors who have made this book possible.

The editors acknowledge the support of the European Commission for funding aspects of the research required for the production of this book and Queen Mary, University of London, U.K for providing an excellent working environment. The editors would also like to Dr John Bigham for help with providing valuable feedback on initial versions of this book.

1

Agent Technology for Communications Infrastructure: An Introduction

S. J. Poslad, R. A. Bourne, A. L. G. Hayzelden, P. Buckle

1.1 Introduction

This introductory chapter outlines the increasing need for software support in the design and implementation of the rapidly expanding communications industry and argues that the agent-based computing paradigm is the most promising approach with which to address these issues. Communication systems are expanding both in physical terms, with the exponential growth in internet usage and the mobile phone market, and in the range and complexity of services offered by providers. Recent developments in wireless application programmes (WAP) which are bringing these two technologies closer together only suggest that larger and more sophisticated software systems will be needed to cater for the increasing demand. Approaches that facilitate the design and implementation of these new communication systems are therefore required. Agent-based computing, and in particular research into multi-agent systems, is leading the way in the development of techniques to address the variety of needs associated with these new technologies: languages and protocols which enable autonomous systems to communicate, frameworks for negotiation between self-interested parties, platforms which allow mobile processes to operate at remote sites, and many more.

To provide the interested reader with both the background motivation and an introduction to the software agent philosophy, this chapter outlines the reasons underlying the need for more radical approaches to communication systems design (section 1.2) and the agent-based computing paradigm as a design methodology (section 1.3). For readers more familiar with these issues, section 1.4 reflects the authors' views on the current state-of-the-art in agent technology with pointers to recent research results in its application to the communications industry. In section 1.5, we overview the research papers brought together in this volume which, though not comprehensive, provide a snapshot of both the variety of problems which can be addressed using software agents and the novel ways in which researchers are attacking them.

1.2   Communications Infrastructure for Next-Generation Services

1.2.1   Mass production for a mass consumption service infrastructure

The start of the electronic communication and information service industry was dominated by a relatively small number of powerful service providers and service delivery channels. These faced little or no competition, were heavily regulated and licensed, and offered a relatively fixed product and service range. The product and service requirements were supplier-driven, determined and fixed by mass production, and they supported limited customisation by the end-user. A poor match between the user requirements, perceived by the suppliers before production and the actual user requirements impaired service uptake. Customers often needed to modify their business models and requirements substantially to fit them to the information and communication products on offer (McDermid, 1998).

Under pressure from different market forces such as increasing privatisation, deregulation, diversification and competition, the customer beholds increasing choice, complexity and heterogeneity at a variety of infrastructure levels including types of message transport (e.g. wired-voice, wireless-voice, wired IP), portal, service and customer interfaces. The communications and information industry along with other types of industry is evolving from a predominantly mass production model towards a model that also incorporates a mass customisation service infrastructure.

1.2.2   Consumer-driven customisable service infrastructure

Customisation can be defined as the ability of a product or service to be modified and maintained to meet particular customer requirements or profiles. Customisation can occur at multiple levels and can be customer, third-party (e.g. broker) or provider driven. Here customer (or consumer), broker, and provider, are just temporal roles played depending on an entity's position in the supply demand chain for a particular service. Mass customisation can be considered as a form of mass production, differing from it only in terms of granularity and abstraction. For example, in the late 1990s, customers became able to switch to an additional service provider for cheaper international voice calls by dialling additional access numbers at the start of a call; at this level, customisation is customer-driven. A new communication service can be synthesised from combinations of services from different vendors, resulting in a cost reduction for both long-distance and other calls.

However, the complexity of integration of multiple service components may overwhelm the consumer. For example, service providers may appear or disappear overnight; lucrative service opportunities may be temporal in nature; and integration may require specialist knowledge that is not available to the customer. This seeds a growth in third-party intervention, such as human and electronic service brokerage, offering independent advice to users to help them intelligently search and select service combinations. A broker may be able to simplify, configure and coalesce access to multiple services for the user. For example, TV set-top boxes, telephones and PCs can all behave as a service portal to integrate information from the internet, voice conversations and broadcast video.

For optimal customisation, service providers need to produce interfaces and delivery channels with suitable designs and abstractions to support reconfiguration and cooperation. For example, although customers may simultaneously hold several contracts on different tariffs between multiple service providers, there is no easy way to produce an integrated service bill – this may require existing services and channels to be re-engineered or wrapped.

Providers or third parties who wish to customise or slant services and products for selected customers (e.g. through unicast or multicast rather than broadcast), will require detailed and up-to-date access to personal and group customer profiles. In contrast, for customers to customise services requires timely access to service and product configuration information. Due to the potential temporal nature of both providers and their products and services and the opportunity to re-evaluate contracts when desired, automated processing of these requirements becomes expedient.

1.2.3   User requirements for a highly customisable service infrastructure

We now discuss several specific user requirements that support a highly customisable service infrastructure, extending the generic requirements of distributed systems such as openness and scalability. These comprise:

- Accessible, configurable and secure service profiles
- Accessible, configurable and secure customer profiles (also called personal profiles)
- Sophisticated advertising and matching of service provision to service requests
- Cooperation, coordination, control and coherence of groups of autonomous service providers and users.

Service profiles are a high-level publicly configurable form of the service interface that is distinct from the internal developer interfaces. The former are abstract and modular enough to be customisable by service users and their proxies (others designated by users to act on their behalf because they provide the necessary know-how). Service profiles define which elements of the server can be modified. Some sort of grand design is clearly necessary to constrain the service as a whole, and to prevent chaotic and improper use of the service. However, a complete grand design is usually not a viable option, especially for extensive, long-lived software, since its future environment may not be known in advance, and fixed designs may alienate users. Instead, designs should support *habitability,* a specialisation of customisation – this is the characteristic of a service that enables users or their proxies to be comfortable and confident in maintaining or modifying that service (Gabriel, 1996). Although service interfaces may be available, they may not be habitable; for example, they may not be configurable by the user, users may not understand how to use them, or they may be too low-level. Habitable services provide interfaces at the correct level of abstraction and modularisation.

Customer profiles that characterise customer preferences need to be defined to enable services to be customised. Specific vendor-driven initiatives such as *directory-enabled networking* (DEN)[1] seem attractive here, but they often lack the appropriate abstractions and open design to support all parts of this requirement. For example, DEN has focused on modelling, storing and retrieving schemas, which relate customer profiles to network service profiles in directory services supporting LDAP (Light Directory Access Protocol) and Microsoft's Active Directory model. DEN primarily focuses on modelling the information network providers' need to exchange in order to provide interoperable network services such as constructing and supporting end-to-end quality of service across multiple providers, rather than on modelling how users can configure network provision. However, DEN does not define security nor does it define a protocol or policies for exchanging this information between users and service providers.

Another user-specific requirement is support for sophisticated advertising and matching of usage and provision. For example, if a service request cannot be satisfied by a single provider then flexibility is required to progress beyond a simple denial of service – in this case a service request may be satisfied by aggregating possible parts of various services from more than one provider. This requires a holistic view of service provision matching which may involve a more complex negotiation protocol to exchange and agree the specifics. We also need to be able to share a common understanding of information across multiple domains. For example, while frame-relay network providers may define quality of service using parameters such as FECNs and BECNs (Forward and Backward Congestion Notification bits), customers may better understand terms such as peak and average throughput, delay and delay jitter, discard/loss rate and overall availability.

Finally, we need groups of services to be controllable, i.e. to be coordinated and cooperative so that the set of services as a whole is coherent. This becomes increasingly necessary when participating services are autonomous, or do not belong to static organisations with a fixed command and control structure. For example, when previously autonomous services are aggregated, cooperation often involves dynamic peer-to-peer interaction to resolve problems such as feature interaction or resource contention over bandwidth.

1.2.4   *Support for next-generation services*

The issues raised above have implications for the development and maintenance of current and future e-commerce and e-business applications and the underlying network infrastructure. All such services will require rapid and flexible implementation of software support. There is clearly a need to ensure that this can take place using robust but adaptable design solutions. We argue herein that the agent-oriented programming paradigm facilitates this process and is therefore the most likely design methodology to fulfil the requirements of the next generation of services.

The research featured in the coming chapters amply demonstrates the flexibility of agent-oriented design when applied to new services such as Virtual Home Environments (VHE) and customisable datacom services, as well as enabling the rapid development of

---

[1] DEN. Available: http://www.dmtf.org/spec/denh.html [18 May 2000].

simulation platforms from which to test the impact of new services. Furthermore, in the area of network management and infrastructure, the use of software agents has led to the development of radical new solutions and capabilities.

Before the reader dives into these exciting new software approaches, there is a need to reflect on the underlying themes of agent-oriented programming and to highlight the reasons that this approach is well-suited to such a diverse problem domain.

1.3   Agent Architectures

At their simplest, software agents are just independently executing programs – often implemented as multiple threads or UNIX-like processes – which are capable of acting autonomously in the presence of expected and unexpected events. Agents can be of differing abilities, but typically possess the required expertise to fulfil their design objectives. To be described as 'intelligent', software agents should also possess the ability of acting autonomously, that is, without direct human input at run-time, and flexibly, that is, being able to balance their reactive behaviour, in response to changes in their environment, with their proactive or goal-directed behaviour; in the context of systems of multiple autonomously acting software agents, they additionally require the ability to communicate with other agents, that is, to be social (Wooldridge, 1999).

The central concept which distinguishes software agents from simple programs is their interaction with their environment; typically, we say that an agent is embedded or situated within its environment (Jennings and Wooldridge, 1998). While it is true that any distributed system could be implemented as a single centralised system, such an approach ignores the fact that, in some types of environment, information and control are naturally distributed throughout the system; this is particularly true of communication systems. Distributing agents to deal with information and control locally, while enabling communication at higher levels of abstraction, can make systems easier to understand and hence make them easier to design and develop.

In this sense, agents can be embedded throughout a system performing functions with appropriate levels of expertise. For example, in chapter 11, it is shown that using a layered hierarchy of agents to manage a network (similar to an organisational structure), enables both long-and short-term decision-making to occur independently, providing robust and rapid responses to short-term fluctuations in network demand, with longer-term considerations such as network survivability being controlled at the higher management plane.

Agents can act as intelligent decision-makers, active resource monitors, wrappers to encapsulate legacy software, mediators, or as simple control functions. The common link between this vast array of possibilities is that each agent autonomously strives to meet its own design objectives. In some ways, agent-based computing can be seen as a natural extension of object-oriented programming whereby objects are empowered with their own thread of control and their own goals or objectives, so that they autonomously control their own behaviour in order to reach their goals. An excellent introduction to intelligent agents and multi-agent systems can be found in Weiss (1999).

1.3.1   Mobile agent systems

The term mobile agent encompasses two distinct concepts: mobility and agency. While some researchers regard the mobility of an agent as its defining feature, with the support for agent behaviour secondary, others regard an agent with mobility to be simply a special type of agent.

Mobility can provide a useful mechanism to enable modification of both services and customer profiles at separate locations, potentially protecting privacy of both parties; subsequently, the updated service or profile can be transferred to the other party for integration. However, additional abstractions are required, such as agency, to support advertising and matching of service usage to service provision, and to support cooperation, coordination, control and coherence of providers and customers. Generally, the mobile agent frameworks or tools mean that there is a large platform overhead for many problem domains – the associated overhead of providing a mobile agent solution may outweigh its benefits in some domains.

1.3.2   Multi-agent systems

The ability of an agent to be social and to interact with other agents means that many systems can be viewed as *multi-agent systems* (MAS). Our definition of software agents as autonomous processes means that they encapsulate their own state and behaviour. For agents to interact, they must also possess the ability to communicate with other agents via some common *agent communication language* (ACL). Similarly, for agents to cooperate with one another or to coordinate activities, requires some inherent desire to be social, whether this is due to self-interested, utility-maximising criteria among competitive agents, or because the system as a whole is designed to achieve some ultimate goal, through essentially cooperative agents pursuing the common good. Any such MAS must include infrastructure to support transport and directory services among agents, ACL message encoders and parsers, dialogue management, and brokers and facilitators.

Multi-agent systems can adopt a range of sophisticated approaches for advertising services, customer preferences and subsequently matching service usage to service provision through interaction between service, user and facilitator agents. Agents can deploy a variety of strategies for the cooperation, coordination, control and coherence of groups of services and users. For example, service and user agents can send messages within the context of interaction sequences such as client server interactions, auction mechanisms, the contract net protocol, and different types of brokerage to support complex negotiation strategies. Multiple providers and users who share common service and customer ontologies have a basis to gain a much deeper agreement in the provision and use of services.

1.4   Agent Technologies

Having made a case for exploiting agent and multi-agent system designs to develop next-generation services and infrastructure, we now turn to the state-of-the-art support for agent technology. This section describes the current status of agent standards, communication languages and toolkits which support development of both open and dedicated multi-agent systems. The FIPA agent standards and the agent toolkit FIPA-OS are then looked at in more detail.

1.4.1   Agent standards

Early adopters of new technology tend to be wary when there are no commonly agreed, well-defined interfaces that are also backed by industry-supported standards. The standardisation process shifts the emphasis from longer-term research issues to the practicalities of realising commercial systems. For this reason, there has been a growing force behind developing agent standards. Such standards provide added confidence to potential adopters of agent technology. However, full compliance to complex agent standards can lead to large overheads in terms of time-to-market and adherence to unnecessary features. Moreover, if standards bodies are not careful, users will not take on board this concept – speed, efficiency and robustness are still top priorities in the communications industry.

Agent standards bodies have reflected the distinction between stationary and mobile agents: the Foundation for Intelligent Physical Agents (FIPA)[2] and the Object Management Group's Mobile Agent System Interoperability Facility (MASIF).[3]

FIPA is a non-profit standards organisation established in 1996 to promote the development of specifications of generic agent technologies that maximise interoperability within and across agent-based applications. Part of its function is to produce a specification for an agent-enabling software framework. Contributors are free to produce their own implementations of this software framework as long as its construction and operation comply with the published FIPA specification. In this way, the individual software frameworks are interoperable. The FIPA agent standard aims to bring the commercial world a step closer to true software components, the benefits of which will include increased reuse, together with ease of upgrade. FIPA allows for focused collaboration (of both industrial and academic organisations) in addressing the key challenges facing commercial agent developers as they turn agent technology into products.

OMG's MASIF differs from FIPA in that it regards the mobility of an agent from one location to another to be its defining characteristic. MASIF defines interaction in terms of Remote Procedure Calls (RPC) or Remote Method Invocation (RMI); it does not support or standardise interoperability between non-mobile agents on different agent platforms. Further, MASIF restricts the interoperability of agents to agents developed on a single type of platform (e.g. agents deployed on one Grasshopper platform can migrate to another Grasshopper platform) whereas the focus of FIPA is to directly support the interoperability of agents deployed on heterogeneous agent frameworks.

Stationary and mobile agent standards are, however, starting to overlap and may ultimately converge. There is strong interest within FIPA, particularly by wireless network operators and service providers, to extend its approach to support mobile agents.

---

[2] FIPA. Available: http://www.fipa.org [18 May 2000].

[3] MASIF. Available: http://www.fokus.gmd.de/research/cc/ecco/masif/index.html [18 May 2000].

Meanwhile, the Object Management Group is exploring the development of standards for stationary agents in addition to mobile agents, and FIPA is inputting into this process.

1.4.2 Agent communication languages

Communication enables agents to coordinate their actions and behaviour, resulting in systems that are more coherent; through coordination, agents can better achieve their design objectives (their own or the system's goals). Whether agents inhabit a dedicated system, or an open one, some means of communication among agents needs to be supported. Long before the drive towards agent standards was initiated, attempts at standardising agent communication languages was well under way.

Communication among agents has most successfully been modelled using *speech act theory* (Searle, 1969). Speech act theory uses the term *performative* to identify the intention behind a spoken communication; examples include verbs such as request, tell, report, commit or reply. These performatives are used to constrain the semantics of each act of communication among agents, so simplifying how agents should react to messages they receive.

Knowledge Query Meta-Language, or KQML (Finin and Fritzson, 1994), was one of the first initiatives to specify how to support the social interaction characteristics of agents using a protocol based on speech acts. However, KQML is not a true de facto standard since there is currently no consensus in the community on a single specification (or set of specifications), nor has it been ratified by common agreement between members of an organisation or forum of some standing in the community. As a result, variations of KQML exist and different agent systems may speak different dialects which prevents full interoperability between them.

Any act of communication involves three aspects: the method of message passing, the format, or syntax, of the information being transferred, and the meaning, or semantics, of the information (and message). The syntax of a generic message in KQML is given in Figure 1.1.

```
(KQML-performative
        :sender    <word>
        :receiver  <word>
        :language  <word>
        :ontology  <word>
        :content     <expression>
        ...)
```

Figure 1.1   The syntax of a generic KQML message

One of the main problems with KQML, notwithstanding the lack of an agreed specification, is its lack of a well-defined semantics. The use of performatives alone is insufficient to guarantee that messages will be interpreted and acted on correctly by other agents.

This lack of semantics is one of the driving forces behind the FIPA specification of its own agent communication language (ACL) which provides an attempt at a universal message-oriented communication language (FIPA, 1997). The FIPA ACL describes a standard way to package messages, in such a way that it is clear to other compliant agents what the purpose of the communication is. Although there are several hundred verbs in English corresponding to performatives, the FIPA ACL defines what is considered to be a minimal set for agent communication (it consists of approximately 20 performatives). This method provides for a flexible approach for communication between software entities exhibiting benefits including:

- Dynamic introduction and removal of services
- Customised services can be introduced without a requirement to recompile the code of the clients at run-time
- Allowance for more decentralised peer-to-peer realisation of software
- A universal message-based language approach providing a consistent speech-act-based interface throughout software (flat hierarchy of interfaces)
- Asynchronous message-based interaction between entities.

Another attempt at providing a common language among agents and legacy software, is the Knowledge Interchange Format or KIF (Huhns and Stephens, 1999). This is based on first-order logic which is capable of describing almost any concept of interest or utility to people and other intelligent agents. Agents which can translate to and from KIF have the potential to exchange information with any other KIF-speaking agent or software system.

Even if agents speak the same language, they require some common understanding of the meaning of the message content. This is provided by a specialised knowledge component called an *ontology* which specifies the objects, concepts and relationships in a given domain (Guarino, 1995). Agents which share a common language and an ontology can communicate effectively both syntactically and semantically.

1.4.3   Agent toolkits

For agent technology to fulfil its promise, agent toolkits need to become widely available and must be robust enough to be used in e-business sectors such as telecommunications. These tools must be able to be applied to address the customer requirements and ideally they need to adhere to standards which define multi-party agent interoperability.

Currently, a profusion of agent toolkits (see Table 1.1) are being made available for use by third-party developers, application developers and end-users under licensing arrangements which range from open source to commercial. Agent tools and systems have mainly been developed within research laboratories funded by governments; at this time, they are being deployed to develop and test concept demonstrators and very specialised one-off applications for a new generation of communications services, rather than to underpin core business solutions or mass-production end-user applications. This new generation of communication services include: Virtual Home Environments, Virtual Private Networks, video conferencing, etc.

For agent toolkits to be taken up by application developers, moving out of the laboratory and into more mainstream development, they need to:

- Provide suitable abstractions, interfaces
- Adhere to appropriate and proven industry standards
- Generate software that can be embedded within a non-agent software infrastructure.

Agent toolkits are defined as *sets of components* from which to build agent systems and *sets of tools* to help operate agent systems. The agent system is somewhat different from the agent toolkit itself in that when it executes, it represents one or more particular configuration(s) of the toolkit.

The first generation of multi-agent systems, which derived from Distributed Artificial Intelligence (DAI) in the 1980s, consisted of components which were tightly integrated, often used a proprietary, shared memory-data model to communicate, and were implemented using specialised languages, often with proprietary extensions. The concept of agent toolkits emerged in the next generation of agent systems in the 1990s (see Table 1.1), and used standard message-passing protocols to communicate knowledge or to migrate the agent itself to a new location. These were implemented using mainstream languages such as C++ or, more commonly, Java. Common functions such as agent-level communication, message transport and directory service were encapsulated into components with well-defined interfaces.

Table 1.1   Publicly available agent toolkits

Product Name (Company) URL	Mobile / Stationary	Language	Standards	Availability/ Licensing	Example applications
AgentBuilder (Reticular Systems) http://www.agentbuilder.com	Stationary Mobile	Java		Evaluation, Academic	E-commerce, auctions, job-finder, private e-mail, etc.
Aglets (IBM Japan) http://www.trl.ibm.co.jp/aglets	Mobile	Java	MASIF	Evaluation	Remote Monitoring (e.g., Web site), meeting scheduling, auctions, etc.
APRIL (Fujitsu, USA) http://www.nar.fujitsulabs.com		APRIL	FIPA	Open Source	Schedule Mgt, etc.
Comtec Agent platform (Japan) http://www.fipa.org/glointe.htm	Stationary	Java	FIPA	Open Source	Enterprise Information Mgt.
Concordia (Mitsubishi) http://www.meitca.com/HSL/ Projects/Concordia	Mobile	Java		Evaluation	Information retrieval, etc.
DMARS (AAII) http://www.aaii.oz.au/proj/ dMARS-prod-brief.html	Stationary	C/C++		Superseded by Deimos	Air traffic control, TNM, simulation, fault diagnosis, etc.

Table 1.1 (continued)

FIPA-OS (Nortel Networks) http://www.nortelnetworks.com/ fipa-os	Stationary	Java	FIPA	Open Source	Personalised services, VPN, VHE, meeting scheduler, etc.
Grasshopper (ikv++) http://www.ikv.de/products/ grasshopper/grasshopper.html	Mobile	Java	MASIF/ FIPA	Evaluation	E-commerce, information retrieval, etc.
Jackal (UMBC) http://www.alphaworks.ibm.com/ tech	Stationary	Java	KQML	Evaluation	Enterprise information mgt, manufacturing planning, etc.
JADE (CSELT) http://sharon.cselt.it/projects/jade	Stationary	Java	FIPA	Open Source	Travel assistant, audio-visual entertainment, meeting scheduler, etc.
JAFMAS (Uni. Cincinnati) http://www.ececs.uc.edu/~abaker/ JAFMAS	Stationary	Java	KQML	Unrestricted licence	Supply chain mgt, etc.
JATLite (Stanford Uni.)	Stationary	Java	KQML	Open Source	Design decision tracking, constraint mgt, enterprise control.
Jess (Sandia Nat. Labs) http://herzberg.ca.sandia.gov/jess		Java		Academic	Various.
MOLE (Uni-stuttgart) http://mole.informatik.uni-stuttgart.de/	Mobile	Java		Evaluation, Academic	information retrieval, groupware, active documents, etc.
Open Agent Architecture (SRI) http://www.ai.sri.com/~oaa/main. html	Stationary	Java, C, VB, Lisp, Prolog, etc.	KQML	Academic	Automated office, robot control, collaborative fridge, etc.
Voyager (ObjectSpace) http://www.objectspace.com/	Mobile	Java		Academic	Various.
ZEUS (BT UK) http://www.labs.bt.com/projects/ agents/zeus/	Stationary	Java	FIPA KQML	Open Source	Supply chain mgt, service provisioning, network resource mgt.

Table 1.1 classifies agent toolkits according to whether they support stationary agents (intelligent messenger) or mobile agents (intelligent message), or both. In addition, it highlights the language in which they are implemented, which communication standards they support at the agent level, the type of availability and licensing, and the types of applications or services they have been used to develop.

The availability of toolkits can range from use of the binaries for a fixed evaluation period, to unlimited use for academic (i.e. non-commercial) purposes, to unlimited use of both the binaries and source code for academic use, to a full open source-code licence for academic and commercial use. JATLite was one of the first stationary agent toolkits made publicly available; it later became available as open source. FIPA-OS was the first toolkit available under full open source (at least the first one that supports the FIPA agent standards); it was released by Nortel networks in October 1999. Since that time, others such as BT's ZEUS platform, CSELT's JADE and Comtec's agent platform have followed suit.

Sun's Java is the current hot implementation language for agent toolkits, with one or two exceptions such as APRIL (a language in its own right) and dMARS (implemented in C++). The latter is no longer available, but has been superseded by Deimos which is implemented in Java. Some developers, particularly those in the DAI community, consider an object-oriented language imperative, but languages like Java tend to be fairly static and unsuitable for the kind of dynamic rule and fact manipulation required to support a notion of intelligence. Although new Java objects can be loaded dynamically, this is at a fairly coarse level of granularity and does not have suitable abstractions and formalisms to support sophisticated rule-based, reasoning systems. To support this latter requirement, Jess (Java expert system shell), a rule engine and scripting environment, was developed by Ernest Friedman-Hill, at Sandia National Laboratories in Livermore, CA. Using Jess, you can build applications that have the capacity to 'reason' using knowledge supplied in the form of declarative rules. JESS, which is implemented in Java, is increasingly being used in agent toolkits such as FIPA-OS, JADE and JATLite, and to support complex reasoning. For this reason, we have included JESS in Table 1.1, although strictly speaking it is not an agent toolkit.

There are many multi-agent systems currently being researched, developed and used. The remainder of this section discusses a subset of these, focusing on those which use an agent communication language for communication between multiple agents and which support the FIPA specification, hence promoting openness in terms of extensibility and licensing.

Most agent platforms, including FIPA and non-FIPA platforms, naturally offer openness at the agent level; whether or not the platform itself is fixed or closed, service and user agents can be dynamically added to the platform and can interoperate. This requires a common means of representing (encoding), understanding (ontology) and exchanging (protocol) service information. FIPA platforms define a standard protocol based on speech acts and several standard ontologies: a core one for registering, querying and deregistering services (part of the FIPA management ontology) and various domain-specific ones; there is no mandatory specific service encoding. Non-FIPA platforms require the use of platform-specific service ontology, encoding, and protocol combinations.

Many agent platforms support some type of bootstrap process which includes agent synthesis for resident agents; often an agent shell or agent factory API is defined. The shell contains hooks into the platform to use lower-level services such as a transport service. To synthesise agents, it is not strictly necessary to use the platform API. Providing a standard protocol such as a TCP/IP can be understood by the platform for the exchange of messages, the non-resident agent can be synthesised externally and registered with the platform. Of course, resident and non-resident agents may be managed by the platform differently.

Several non-FIPA platforms such as OAA, JATLite and JIAC offer some degree of openness for reconfiguration, including replacement of the platform services. For example, JIAC describes several interchangeable components including service broker, message transport, knowledge base and security.

There are now several MAS platforms that report support for the FIPA agent standards: JADE, Grasshopper, ZEUS and FIPA-OS. Of these, only FIPA-OS, JADE and ZEUS are freely available under a general public licence and are released with source code under an open source licence.

There are several important differences between FIPA-OS, JADE and ZEUS. FIPA-OS (Poslad et al. 2000) and JADE were designed at the onset to focus on supporting agent communication using the FIPA agent standards, whereas ZEUS initially supported only KQML as the agent communication language and has since been rewritten to support the FIPA ACL. FIPA-OS was designed to operate in a heterogeneous open service environment – it does this by supporting multiple transports such as IIOP using a variety of CORBA APIs, RMI and TCP and by supporting multiple encodings for the content. The interoperability of the publicly released version of FIPA-OS is being evaluated in a heterogeneous FIPA environment in the FACTS project. The current ZEUS platform supports Part 2 of the FIPA ACL, but it does not yet support Part 1 of the FIPA 1997 specification (agent management), nor has its interoperability with non-ZEUS FIPA platforms in a heterogeneous environment yet been evaluated. FIPA-OS initially focuses on providing abstractions and interfaces for developers who wish to extend, enhance and integrate an agent platform with existing software infrastructures. ZEUS provides many higher-level abstractions for developing agents – it hides the API and instead provides an integrated development environment to allow developers to configure the agent platform.

1.4.4   The FIPA agent standards

There are several versions of the specifications for FIPA standards in existence. Currently the most widely implemented version is FIPA 97 Version 2.0. In practice the two core parts are: Part 1, which defines the agent platform or agent reference model, and Part 2, which specifies the agent communication language. The FIPA 97 agent reference model (Figure 1.2) provides the normative framework within which FIPA agents exist and operate. Combined with the agent life cycle, it establishes the logical and temporal contexts for the creation, operation and retirement of agents.

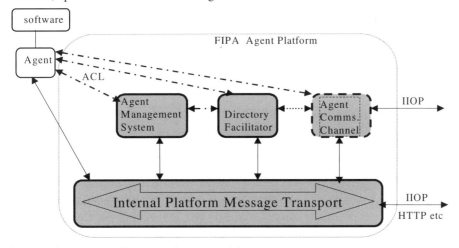

Figure 1.2   The FIPA 97 agent reference model

The agent reference model consists of a Directory Facilitator (DF), Agent Management System (AMS) and Agent Communication Channel (ACC). These are the specific capability sets for agents to support agent management. The DF provides 'yellow pages' services to other agents. The AMS and ACC support interagent communication. The ACC supports interoperability both within and across different platforms. The Internal Platform Message Transport (IPMT) provides a message routing service for agents on a particular platform, which must be reliable and orderly. The ACC, AMS, IPMT and DF form what are also termed the Agent Platform (AP); these are mandatory, normative components of the model. The ACC, AMS and DF are capability sets – they may be performed by a single agent, or three different agents (this is left to the agent platform developer).

To be minimally FIPA-compliant requires compliance to FIPA 97 V2, Part 1, Agent Management, and FIPA 97 V2, Part 2, Agent Communication. There are differing degrees of compliance to, and interpretation of, FIPA compliance. For example, if the FIPA platform has no requirement to support access to non-agent services by agents, then adherence to Part 3 is not required.

The FIPA standards in some areas introduce conceptual problems for designers and implementers. For example, the FIPA Agent Communication Language (FIPA ACL) focuses on an internal mental agency of beliefs, desires and intentions, and closure is not enforced (agents are not compelled to answer); both these concepts hinder multi-agent coordination. The FIPA normative specifications are not intended to be a complete blueprint or specification for building a multi-agent system. For example, FIPA standards do not prescribe how to manage the existential aspects of agents in a discrete world; nor do they define error handling, although some aspects of error reporting are covered. Useful information for developers is also given in an informative part of the FIPA 97 Developers' Guide, Part 13.

The FIPA specifications themselves may evolve to solve current shortcomings and to meet future needs. For example, at present, agents are managed via a message-passing interface at the ACL level, i.e. agents are managed by interaction with the three core FIPA platform agents: the DF agent, the AMS agent and the ACC agent. Currently, an agent often needs to send a forward request to the ACC agent in order to send the payload message. FIPA is currently considering offering one or more of these core services via a non-agent interface such as a method invocation interface rather than by encapsulating it as a service whose sole interface is via the specialised service agent. This may reduce the degree of message passing required during bootstrapping and during the session, thus enhancing scalability. For these reasons, 'FIPA-compliant' agent platforms may also need to evolve to comply with new or replacement standards.

1.4.5   FIPA-OS: an agent platform based on the FIPA standards

FIPA-OS is designed to support the FIPA agent standards. The FIPA reference model discussed earlier defines the core components of the FIPA-OS distribution. Currently, there is no formal or clear mechanism to determine the compliance of a FIPA architecture implementation. Moreover, there are different versions of the architecture, which are not completely backward-compatible. For these reasons, it is only possible to describe which features of a FIPA specification version are incorporated within a given implementation.

At the time of writing, FIPA-OS includes implementations of the core components of the FIPA 97 reference model. In addition, the FIPA-OS distribution includes support for:

- Different types of agent shells for producing agents which can communicate with each other using the FIPA-OS facilities
- Multi-layered support for agent communication
- Message and conversation management
- Dynamic platform configuration to support multiple IPTMs, multiple types of persistence (enabling integration with legacy persistence software)
- Abstract interfaces and software design patterns.

The FIPA-OS architecture can be envisaged as a non-strict layered model, supported by an underlying component model (Figure 1.3). In a non-strict layered model, entities in non-adjacent layers can access each other directly. The developer is able to extend the architecture not only by appending value-added layers, such as specialist service agents or facilitator agents, but in addition, lower or mid-layers can be replaced, modified or deleted.

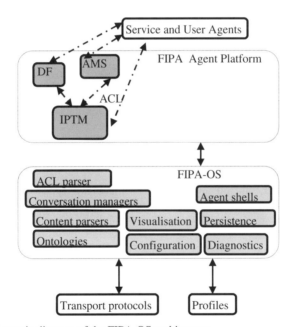

Figure 1.3   A schematic diagram of the FIPA-OS architecture

1.4.5.1   Agent shell

There are a variety of ways in which new agents can be added to FIPA-OS platforms. Agents can be built using different types of agent shell. Agent developers can also create agents without using FIPA-OS services; provided these 'non-resident' agents support a transport protocol supported by FIPA-OS, they can use FIPA-OS to interact with other agents on the platform. Non-resident agents can use or invoke the transport API of the platform directly or use their own non-platform transport.

1.4.5.2   Multi-tiered ACL communication

Understanding an ACL message requires processing the message with regard to its temporal position within a particular interaction sequence between two or more agents. This involves understanding the type of communication called a communication act (which is specified in the message, e.g. a request, statement of fact, or query), understanding the structure of the content and finally understanding the semantics of the request.

As ACL communication is so rich, it is often represented as a multi-tiered layer in its own right. FIPA-OS supports ACL communication using four main sets of components: conversation, ACL message, content (syntax) and ontology (content semantics). Not all of these components need to be used by each agent, and combinations of different types of components can be supported at each layer (see below). This flexibility is needed because, in a heterogeneous world, different agents may encode and transport the content differently.

1.4.5.3   Conversation management

Multiple messages are sent as part of an interaction protocol, for example, the base FIPA multi-agent interaction protocol called FIPA-request supports client server request reply-type dialogues. Unlike typical RPC invocation, requests are indirect. Agent A can request another agent B to perform some action on its behalf; agent B normally sends back an agree or a refuse message as a reply to agent A.

Without conversation or dialogue management, messages are sent and received independently of any context within any interaction pattern or conversation in which they occur. It is more difficult here to detect whether failures such as an absent reply or inappropriate reply occurred.

As an aid to conversation management, FIPA provides two hooks in the ACL itself but the semantics of those hooks are yet to be defined. Firstly, the FIPA ACL defines two fields 'reply-with' and 'in-reply-to'; however, the semantics of how these fields are to be used by the sender and receiver, and how the values of these two fields are related, are as yet undefined. Secondly, the FIPA ACL defines a Conversation-ID field; FIPA-OS defines semantics to use the Conversation-ID field to coordinate conversations.

1.4.5.4   Customisation

There is built-in support to plug in different components at key interfaces or hot spots including: transport, message content encoding and ACL message storage. These are plugged in during platform initialisation. The plug-in nature of the platform supports a dynamic component-oriented middleware model, and promotes a combination of thin client agents and thin platform services. Agents are not required to implement platform services themselves: they access these services in the platform (supporting thin agents), and platforms do not need to load support for all types of each service at start-up (leading to thinner platforms).

The types of these plug-and-play components are currently configured in terms of profiles for the platform as a whole and for each individual agent. These profiles are encoded in XML/RDF and stored in resource files.

1.5   Summary and Following Chapters

This chapter has provided the reader with an introduction to agent technology with a focus on how agents have been applied to service delivery platforms. We have discussed various aspects of agent technology, from the discussion of what is an agent, through the various architectures that designers can use to design and build agent systems, to the latest on how agent standards are evolving. The later sections have led the reader through various detailed applications of agent technology. Chapters 2 to 10 of this book extend these ideas by considering the issues related to service and application support for the enabling of e-business. Chapters 2 and 3 provide the scope of application of agent technology in networked environments and provide an insight into how agents in software terms can be developed for e-business applications. Chapter 4 describes the use of learning agents to capture feedback control for multimedia communications management.

E-business can be described as the process of conducting trading or facilitating traditional commerce activities via the use of the Internet. At the time of writing, using the Internet as a platform to deliver novel services and for conducting business processes has become one of the most discussed topics in the popular press. In this vein, chapter 5 is concerned with how agents may facilitate a variety of relevant e-business functions, such as advertising, matchmaking and brokering; the chapter presents a general survey of key enabling techniques that are needed to build institutions of agent-mediated trading within the Internet. Chapters 6 and 7 focus on the security issues that are intrinsic to networked systems. Chapter 6 relates to the security problems and possible solutions for using agent technology for resource allocation such as deciding on bandwidth utilisation measures. Chapter 7 tackles the concerns about secure transactions for buying and selling products over the Internet with a special focus on how agent technology can provide one such solution. Chapters 8 and 9 offer differing approaches to how multi-agent systems can provide a solution to the problem of delivering information to the mobile user. Chapter 10 continues in a similar application stream and explains the use of algorithms and agent technology for planning and supporting individual travel services.

Chapters 11 onwards constitute the second Part of the book and cover the application of agent technology to network infrastructure. With the rapid development of new and larger communications networks comes an increased need for integration of diverse and distributed underlying hardware and middleware platforms. This need has led to a variety of interesting agent-based solutions, some of which are represented here. Chapters 11, 12, 13 and 14 describe agent architectures that have been used for network management and control in both wired and wireless networks. Chapters 15, 16 and 17 show how distributed monitoring of real-time network parameters by agents has been used for prediction, survivability and optimisation. Chapters 18 and 19 apply market-based control techniques to allow economic agents to manage call routing and load control in both ATM and intelligent networks. Chapter 20 uses a collaborative multi-agent architecture to track and resolve faults in global connections. Chapter 21 shows how agents have been used to deal with distributed search for resources. Chapter 22 enables agents to evolve routing algorithms using genetic programming. While this selection can only provide a snapshot of on-going research in both academia and industry, it is clear that researchers in the network domain have found the conceptual framework of agents and multi-agent systems particularly applicable. As the communications industry continues to expand, we envisage that agent-based techniques will be at the forefront of the research and development required to engineer the network infrastructure for the twenty-first century.

1.6   References

Bellifemine, F., Rimassa, G., and Poggi, A. (1999) JADE - A FIPA-compliant Agent Framework. *Proceedings of PAAM 1999, London.*

Finin, T. and Fritzson, R. (1994) KQML: A Language and Protocol for Knowledge and Information Exchange, in *Proceedings of 19th International Distributed Artificial Intelligence Workshop*, pp. 127–136, Seattle, USA.

FIPA (1997) FIPA 1997 Version 2.0 specifications. Available: http://www.fipa.org/spec/fipa97.html [18 May 2000].

Gabriel, R. (1996) *Patterns of Software: Tales from the Software Community.* Oxford University Press, Oxford.

Guarino, N. (1995) Formal Ontology, Conceptual Analysis and Knowledge Representation. *International Journal of Human-Computer Studies*, **5/6**, 625–640.

Huhns, M.N. and Stephens, L.M. (1999) Multiagent Systems and Societies of Agents, in G. Weiss (Ed.) *Multiagent Systems*, pp. 79–120, MIT Press, Cambridge, MA.

Jennings, N.R. & Wooldridge, M.R. (1998) *Agent Technology Foundations, Applications and Markets.* Springer-Verlag.

McDermid, J. (1998) The cost of COTS. *IEEE Computer*, **31:6**, pp. 46 52.

Poslad, S., Buckle, P. and Hadingham, R.G. (2000) The FIPA-OS agent platform: Open Source for Open Standards, *Proceedings of PAAM 2000*, Manchester, UK, pp. 355 368.

Searle, J.R. (1969) *Speech Acts.* Cambridge University Press Cambridge, UK.

Weiss, G. (Ed.) (1999) *Multiagent Systems.* MIT Press, Cambridge, MA.

Wooldridge, M.R. (1999) Intelligent Agents, in G. Weiss (Ed.) *Multiagent Systems*, pp. 27–77, MIT Press, Cambridge, MA.

2

An Agent-based Platform for Next-Generation IN Services

D. Kerr, R. Evans, J. Hickie, D. O'Sullivan, S. Mathews

2.1   Introduction

The World Wide Web has brought about a revolution in terms of the possibilities for multimedia services. This trend is beginning to impact the previously separate arena of telecoms services, with increasing demand on operators and their vendors to respond to customers' requirements for services that comprise combined datacoms and telecoms capabilities. Current IN services are not sufficient on their own to feed the requirements of the customer for increasingly sophisticated, feature-rich, multimedia Internet services. There is thus a need for platforms to support development and deployment of these services, their composition into tailored user services, and support for user interaction. This has been the primary objective of this most recent phase of the PANI project.

The classical Intelligent Network (Kerr, 1999) view of the service chain deals with Service Creation, Service Activation and Service Delivery. These distinct blocks of the service chain reside largely at the Network Level, where a user will make configuration requests to the network (such as call divert (CD), 3-way calling (3WC), call waiting (CW), etc.) directly, the only user terminal active in the process being the telephone handset.

Network Intelligence (Kerr et al. 1998) services add value to this scenario by introducing intelligence at the user level. This could mean having additional peripheral devices connected to the physical network, or designing personal services around existing network-based ones. These could include intelligent call screening, group scheduling, etc., and focus on tailoring user requirements to specific services. Network Intelligence also looks at modes of service activation for classical services, such as PDA-based activation and web-based activation. It is widely agreed that network battles of the future will be fought on the service level, and not the transport level.

2.2   What is PANI?

PANI is an open, dynamic, service platform that deploys elements of agent technology in the provision of value-added service solutions. PANI provides the foundation for the dynamic and responsible delivery of feature-rich, multimedia Internet services which are user-customisable and adaptable to the user's changing environment (Busioc, Winter and Titmuss, 1995).

Previous work in the area of Intelligent Agent Technology (IAT) (Kerr, 1999; Evans et al. 1999) has identified IN as a target domain where the deployment of IAT is beneficial. This work specifically looked at a number of facets to the delivery of advanced user services, from an architecture to the services themselves and their ability to be self-describing to the service platform, thus enabling their immediate availability to end users. A key ability of the PANI architecture is in the composition of basic services into service packages. As example, consider where a user wishes to be notified of important e-mails while travelling. If a service is registered which can deliver information about the users' e-mails and another service registers stating that it can make telephone calls and synthesise speech, then the users can themselves compose a service whereby e-mails are delivered to the telephony system under their required conditions, e.g. high-priority e-mails, or those from a specific person. This approach can be taken further with services composed from a combination of information sources. Basic weather and timer services could easily be combined to create a weekday wake-up call service that triggers 30 minutes early if the weather is wet (as commute times are likely to be longer).

This type of flexibility in service composition and the ability for external services to register themselves and their capabilities makes PANI a unique system. The use of Intelligent Agents allows for powerful rule analysis to be performed and for rules to be fired producing messages to gateway agents such as those responsible for the delivery of voicemails or fax messages.

2.3   PANI Services

A PANI service can be described as being an event service (providing sensor data) or an action service (providing activator data). Sensor data provides the central RulesAgent with the data which is entered to its fact base for processing. Sensor data can be further subdivided into two classes: those which are specific to a particular user (e.g. incoming e-mail messages) and those which are generic to the user population (e.g. stock ticker information). The fact base is in a proprietary format (Friedman-Hill, 1999), and therefore to allow for portability across rule systems a fact meta-language was defined. The services themselves are atomic in that they provide a single function (stock information, GSM SMS message transmission). The user provides the conditions under which the services are combined.

2.3.1   Components of the PANI service architecture

Services are one of two basic types. Event (source) services provide data into the system and action (sink) services provide mechanisms of notification for the user. The events from one service are joined to another service by the RulesAgent, which contains user-specified rules for their connection. One simple configuration of a PANI platform would have a single event service, e.g. stock quote provider, a single RulesAgent and a single action service, e.g. an SMS sending service. The user would configure a rule to send stock quotes pertaining to the Ericsson B share on the Stockholm stock exchange to the SMS service when a condition is met, e.g. the price exceeds SEK300. The service is made available through its registration with the RulesAgent in SDML, the user's GUI renders itself based on available event and action services, and the user composes the rule with the RulesWizard client which pushes SCML into the RulesAgent for verification and activation. The next section will go through this process in more detail.

2.4   Dynamic Service Provision – Service Registration

In order for services to be made available to users dynamically, a process of service registration must be made with the RulesAgent of the system. This is achieved using the meta-language developed – Service Description Meta-Language (SDML). The service registering on the service platform interrogates the platform for an agent that has registered itself as having rules processing capabilities. After a name has been obtained from the platform the agent sends its registration message in SDML. This has the general format and specific example for a news service described in Figure 2.1 below.

```
:tell
:ontology info_service_registration
:language pani_sdl
:content(<service_name>(<param1_name>
(<param1_type>), … , <paramn_name>
(<paramn_type>)))

:tell
:ontology info_service_registration
:language pani_sdl
:content(Deliver_News_Article
(Headline(string), Article(istring)))
```

Figure 2.1 Service registration using SDML

Once the RulesAgent has this message, the service can be made available to the users.

2.5   Service Description and Composition Languages: SDML and SCML

SDML serves two purposes in the PANI system. First, it enables services to be registered on the service network and secondly it allows the user interface to render itself according to the specification of the service and to constrain the user in his/her generation of SCML statements. The dynamic and open nature of the PANI platform place additional requirements on the generation of SCML statements as they are immediately published in the RulesAgent and activated. The client must generate correct SCML statements and the user must not be able to modify the generated statements as they are published in a live system. A comprehensive RulesWizard was developed in order to structure the way in which users create rules and publish these rules to the RulesAgent. The core of the RulesWizard is an SDML/SCML parser/generator that analyses the registrations of the active services and renders GUI widgets appropriate for them 'on-the-fly'. In this manner, the RulesWizard is itself a dynamic component in that it has not got a hard form, but rather a soft one which changes according to the state of the system at the time of client connection.

The range of services which can be implemented is wide and includes all those services which can be described by SDML. Tests have been done to implement varied services in both the datacomm and telecom domains, ranging from delivery of news information (as illustrated in Figure 2.1) to the delivery of alarms in a telecommunications network. Action services in both domains have also been implemented, ranging from sending of e-mails to speech-synthesis over a PSTN network. The core of the service discovery is thus in the SDML basic types supported. To date, several basic types are supported with a view to building more complex types as required.

TYPE	RANGE
integer	-maxinteger to +maxinteger
real	-maxreal to +maxreal
atom[]	list of symbolic terms explicitly enumerated
string	any character string
sstring	any user-searchable character string

Figure 2.2 Basic types offered through SDML

These basic types offer a number of different operators to the user (Figure 2.2), and accordingly to the GUI, which the user manipulates in order to build a service. For example, if a service offers a parameter of type real, the GUI will first render the available operators for that type in a dropdown box (i.e. <, =, >) and then a textbox into which the user will type the operator value. In turn, the RulesAgent will dynamically offer search capabilities to the instances of the service, e.g. a stock value.

2.6   Services Offered in the PANI Prototype

In order to demonstrate the technology developed, a trial system was chosen which would enable users of the trial to configure their own services from the basic services available. The services developed for the trial were:

Event Services:

- News Story – slashdot.org, Dagens Nyheter (with kind permission)
- Stock Quote – NASDAQ, AMEX, NYSE, and Stockholm Stock Exchange
- iMail – incoming mail event service (restricted users only) (Microsoft Outlook)

Action Services:

- Telephony – including dynamic text-to-speech synthesis (Rhetorex) (Microsoft SDK)
- SMS sending
- E-mail sending

A comprehensive GUI was developed in three forms:

- Stand-alone Java application (fat-client approach)
- Servlet-based browser application (thin-client approach)
- A WAP interface for mobile devices such as WAP-enabled phones

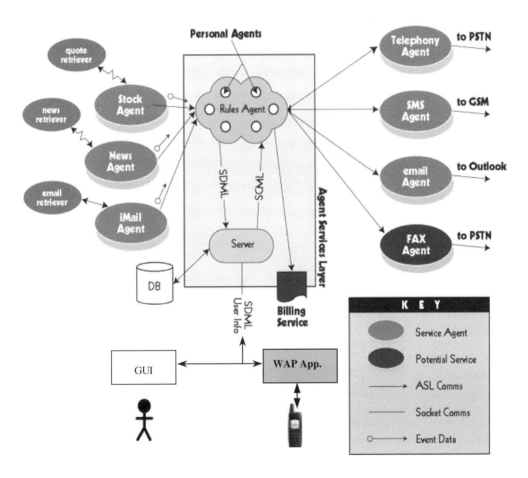

Figure 2.3 PANI trial system

A trial was conducted from March 10th 1999 until March 31st 1999 with the trial participants drawn from both Broadcom and Ericsson P-VAS.

2.7   PANI Services – Implementation

Each of the services provided by the system is developed using the PANI Service API. This is an OO API (implemented in Java) which can be used by platform service developers and third-party service developers alike. It is designed to enable non-agent developers with the tools necessary to deploy their services on the ASL in the PANI architecture. The API consists of a number of objects which represent the service in the PANI service platform and a number of utilities which enable the services to discover and register with the relevant RulesAgent on the system.

```
public class TimeProcessor extends EventServiceProcessor
implements TimeConsumer
{
    private final String CURRENT_TIME = new
String("Current_Time");

    public TimeProcessor() {
       super();
       SERVICE_NAME = "Time";
       serviceDescription = new
       ServiceDescription(SERVICE_NAME);
       serviceDescription.addServiceComponent(CURRENT_TIME,
       ServicePrimitive.STRING);
       TimeScheduler scheduler = new TimeScheduler(this);
       scheduler.start();
    }

    public void updateTime()   {
       sendEventUpdateNotify();
       String[] timeInstance = new String[1];
       java.text.SimpleDateFormat sdf =
       new
       java.text.SimpleDateFormat("HHmm",java.util.Locale.UK);
       java.util.Date current_date = new java.util.Date();
       String current_time = sdf.format(current_date);
       timeInstance[0] = current_time;

       try {
            sendEventUpdate(timeInstance);
          }
       catch (EventUpdateException e)   {
            System.out.println("[ EventUpdateException: "
            + e + " ]");
          } }}
```

Figure 2.4 Complete implementation of a time service on the PANI platform

Broadly speaking, each service will follow the same procedure when deploying in the PANI service platform. These are:

- Discover a RulesAgent
- Construct an SDML expression describing their service
- Send this expression to the RulesAgent

In addition to these three steps, one more step is required and is dependent on whether the service is an event service or an action service. Event services typically inject data into the system for processing and action services typically have data injected into them for processing. Therefore most of the work done by these services is in dealing with either incoming or outgoing data, and the API reflects this.

Objects exist in the API that allows services of each type to have these tasks automated. For example, action services will have all KQML messages automatically parsed for them before the parameters to the services are presented to the service. Likewise, event services will have their data packaged into appropriate messages that are dispatched to the RulesAgent for processing. This automation of service functions enables services to be deployed in the PANI service architecture very rapidly, as illustrated in Figure 2.3. Once this service is deployed, the service will update the RulesAgent with the current time every minute (60000 ms). The next section deals with the user GUI and the interaction with services. Two of the three clients developed will be discussed.

2.7.1   Client interaction

2.7.1.1   Fat-client

The GUI incorporates an SDML parser and an SCML generator. These are used in order to generate widgets required for the user to interact with the services that are currently available on the PANI service platform. The GUI connects to the server using TCP/IP and obtains a list of all available services. Then the RulesWizard is ready to start processing user rules. The first step in setting up a new rule is to select the event service that provides the data to which the user attaches a filter (see Figure 2.5). The user can select a number of heads (conditions) to the rule and can combine events from a number of different services. For example, the user may with to be notified of a stock event to their mobile telephone after working hours and to their desk telephone during working hours, and therefore would combine the stock service wish the time service to achieve this.

After selecting the event and applying a filter, the user must select a notification method (or methods). Each notification method has a number of different parameters that need to be filled in. In the case where the event service supplies a hidden field (e.g. a news article to which the user cannot apply a filter), then this hidden field is automatically filled into a parameter of the action service, as depicted in Figure 2.6. In the case where the event service does not supply a hidden field to the action service, the user must supply all parameters.

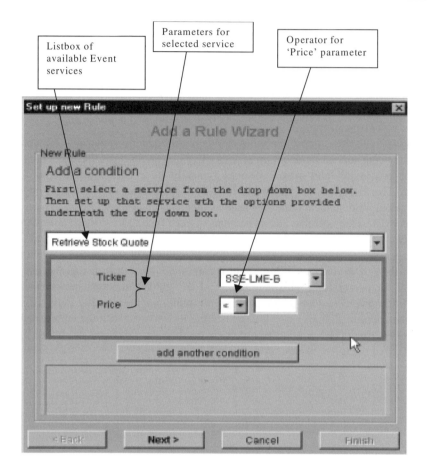

Figure 2.5 Selecting event condition for a new rule

Once the rule is submitted to the RulesAgent in SCML, it is guaranteed by the RulesWizard to be syntactically correct and so can be activated immediately in the RulesAgent. The SCML is converted into a programmatic representation and is asserted into the RulesAgent's rulebase and activated. The first instance of a match to the rule will trigger the firing of the required notification method (SMS, e-mail, telephony). Therefore the user is required to exercise restraint when entering rules so that they have some actual meaning (e.g. a rule to notify the user if the share price of the Ericsson B share exceeds SEK10 would not be as useful as one where notification occurs if the Ericsson B share exceeds SEK300.

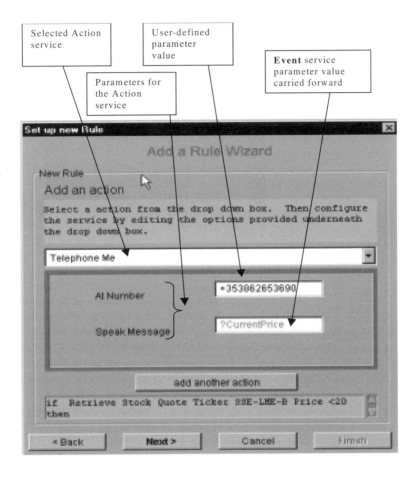

Figure 2.6 Selecting action condition for a new rule

2.7.2 PANI WAP client

The Wireless Application Protocol (WAP) is an application environment and set of communication protocols designed to allow the development of applications and services that operate over wireless networks. Interoperability is an important aspect of the bearer-independent WAP standard. Wireless devices from different manufacturers can all use the services offered by wireless network operators. WAP makes use of existing Internet standards such as XML, URLs, scripting, CGI, content typing, etc. It also utilises mobile networking standards such as GSM, IS-136, etc.

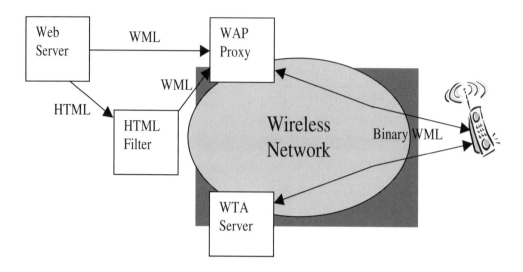

Figure 2.7 An example WAP network

One of the most prominent applications of WAP is to facilitate web browsing through hand-held devices. Figure 2.7 illustrates an example WAP network that would allow the realisation of this application. While a discussion of this WAP application will not be exhaustive it will introduce some aspects of WAP development that are relevant to this document.

Some Internet standards are too inefficient over mobile networks to be integrated into the WAP standard. Examples of such standards are HTML, HTTP, TCP, etc. These standards require large amounts of text-based data to be sent. This is not appropriate for WAP due to the low bandwidth and high latency characteristics of mobile networks.

Another problem is that the small screens of pocket-size mobile devices cannot display HTML content. The solution to these difficulties is WML (Wireless Markup Language) and WMLS (Wireless Markup Language Script). WML caters for displaying content, both text and images for more advanced wireless devices, on small screens. Binary transmission is used to transfer WML pages (WML is actually organised as WML decks which contain WML cards). This allows for data compression and optimised transmission over wireless networks. HTML filters can be used to translate HTML to WML. Wireless Telephony Application (WTA) servers allow the mobile device access to services offered by the wireless network operator.

2.8   PANI and WAP

The integration of WAP with PANI greatly increases the versatility, availability, value and commercial viability of the PANI project. The user's ability to use PANI services without having to be sitting at, or even owning, a computer adds to the attractiveness of the PANI concept. The mobile device is an adequate interface to PANI.

2.8.1 The WAP application

The features of WAP necessary to implement the application to be integrated with PANI will now be discussed.

WML data is structured as a collection of cards. A single collection of cards is referred to as a WML deck. Each card contains structured content and navigation specifications. Logically, a user navigates through a series of cards, reviews the contents (both text and images can be supported) of each card, enters requested information, makes choices and navigates to another card or deck or returns to a previously visited card or deck. The mobile device maintains a certain amount of state information. This information is the browser context and includes navigational history for efficient backward navigation and a number of variables and their values used within WML cards. The browser context can be initialised by the new context attribute of the card element.

Input from the user can be entered as text or the user can choose from a presented list of options. Navigation is event-driven. For example, when the user presses the OK button on a mobile phone an ACCEPT event is generated. This event is caught and triggers navigation to a new card or deck. Decks and cards are identified by URLs.

The Common Gateway Interface (CGI) is a method of allowing a URL to refer to a program rather than a static WML deck or card. If a URL in WML points to a CGI then the server hosting the CGI program will execute it. It is possible to pass variables contained in the browser context to the CGI program as part of the URL. The CGI program must create decks containing cards in order to display its output on the mobile device.

This is all the information that is needed to collect user input, process it and display results on the user's mobile phone. The CGI program was implemented as a perl script. This was because there were modules available for the simple and efficient creation and display of WML decks and cards. It was decided the perl script would pass the information to a Java application. This application would do the following.

- Take input from the CGI program.
- Formulate and send a request to the PANI server using this input, e.g. adding or deleting a rule.
- Extract the relevant information from replies from the PANI server.
- Return this information to the CGI program.
- Dynamically generate WML decks of cards displaying the services offered by PANI, the existing rules for the user, etc.

The perl script then creates the necessary deck of WML cards and displays these on the user's mobile phone. Sockets were used as the communication mechanism between the CGI and Java program. Figure 2.8 details a basic overview of the WAP application.

With the incorporation of the Java program it is possible to remove much of the complexity from the mobile phone. It would be possible to maintain the Java program on a WTA (where the WML cards and decks and the CGI programs would more than likely be hosted). The Java application could update the WML decks as services offered by PANI are added or deleted.

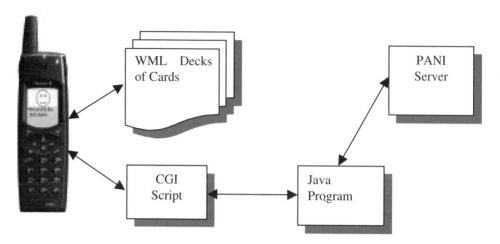

Figure 2.8 WAP application architecture

2.9   Conclusions

The worldwide market for specialised and personalised information services is catered for increasingly by services based both on IN architectures and on the World Wide Web. This trend is set to accelerate, with important growth areas being in services delivered over mobile computing platforms and digital set-top boxes.

The PANI project described in this chapter addresses the issues of allowing users and third-party developers the maximum flexibility in creating and customising information services for delivery using a range of telephony media. The key component in the PANI architecture is the use of a meta-language to support creation of service components that are self-describing. This enables naïve users to plug together and customise their own services using service components supplied by third-party developers. The approach is particularly applicable for deploying services for mobile users. This chapter has described the PANI system in terms of the problem domain, existing solutions, the approach adopted and the demonstrator system which has undergone internal trials.

Acknowledgements

This work was kindly sponsored by ERICSSON P-VAS.

2.10   References

Busioc, M., Winter, C., and Titmuss, R., (1995) Distributed Intelligent Agents for Service Management. In IJCAI '95 Workshop, Montreal, Canada.

Evans, R., Somers, F., Kerr, D. and O'Sullivan, D. (1999) A Multi-Agent System Architecture for Scalable Management of High Performance Networks: Applying Arms Length Autonomy, in

Software Agents for Future Communications Systems, Hayzelden, A.L.G. and Bigham, J. (eds.), Springer Verlag, ISBN 3-540-65578-6.

Friedman-Hill E. J. (1999) JESS, the Java Expert System Shell (Version 5.0a6), Internal Report SAND98-8206 (revised), *Distributed Computing Systems*, Sandia National Laboratories, Livermore, CA, 30 June.

Kerr, D. (1999) PANI: Personal Agents for Network Intelligence™, *Communicate*, Vol. **4(2)**, pp 4-15, January. http://www.broadcom.ie/communicate/vol4/iss2/pani.phtml

Kerr, D., O'Sullivan, D., Evans, R., Richardson, R. and Somers F. (1998) Experiences Using Intelligent Agent Technologies as a Unifying Approach to Network Management, Service Management and Service Delivery, in *proceedings of IS&N98* (http://www.springer-ny.com/catalog/np/jul98np/3-540-64598-5.html).

Microsoft Outlook '98, http://www.microsoft.com/outlook/

Microsoft Research Speech SDK, http://research.microsoft.com/stg/

Rhetorex home page (Lucent Technologies) http://www.rhetorex.com/

Wireless Application Protocol (WAP) Tutorial. http://www.webproforum.com/wap/index.html

2.11   Further Reading

Schneiderman, B., Direct Manipulation: A Step Beyond Programming Languages, *IEEE Computing*. **16(8)** (Aug. 1988), 57-69.

Wireless Markup Language Specification. http://www1.wapforum.org/tech/terms.asp?doc=SPEC-WML-19990616.pdf

3

Java Framework for Negotiating Management Agents

O. Krone, B. T. Messmer, H. Almiladi, T. Curran

3.1　Introduction

The ability to rapidly create and deploy new and novel services in response to market demands is one of the key factors in determining the success of the future service provider. As the high-speed switching and communication infrastructure improves and bandwidth becomes a commodity, it is envisioned that the competition for product differentiation will increasingly depend on the level of sophistication, degree of flexibility and speed of deployment of services that a future provider can offer. These factors in turn depend heavily on the flexibility of the software architecture in place in a provider's operational infrastructure.

The current generation of telecommunication networks is based on architecture over 30 years of age. The primary deficiency of this architecture is its monolithic view of the service provisioning process. Interfaces to the service management infrastructure are often non-existent, proprietary or narrow in scope and intimately coupled to the hardware they operate on. As a result, any third-party involvement in service programming is limited to customising only a small set of operational parameters. Furthermore, deploying new services in today's telephone networks takes up to several years primarily because the software systems with which they need to be integrated are enormously complex and prone to many cross-service interactions.

A new paradigm, agent technology, now opens the opportunity to tackle many of the above mentioned problems in customising large monolithic software chunks and badly documented interfaces. A real decoupling of telecommunications hardware infrastructure from its software applications can be achieved with agent technology. In this chapter, an approach to using agent technology in the network management domain based on negotiating agents is presented. In particular, the framework on which the so-called Java management agents JAM are built is described. Furthermore, some real-world applications that have been strongly inspired by the concept of JAM are introduced.

The rest of this chapter is organised as follows: section 3.2 gives a short introduction to agent theory as a whole, followed by section 3.3, which introduces our platform. Section 3.5 is devoted to a short overview of application areas where our platform has been used, and section 3.6 finally concludes this chapter.

3.2   Preliminaries

3.2.1   What is an agent?

Avoiding a new definition and being conscious of the difficulty of the task because of the great variety of existing agent paradigms, we propose to adopt a definition proposed by Wooldridge and Jennings (Wooldridge and Jennings, 1995) who define agents as a hardware- or (more usually) software-based computer system that enjoys the following properties:

- *autonomy*: agents operate without the direct intervention of humans or others, and have some kind of control over their actions and internal state;
- *social ability*: agents interact with other agents (and possibly humans) via some kind of agent-communication language;
- *reactivity*: agents perceive their environment (which may be the physical world, a user via a graphical user interface, a collection of other agents, the Internet, or perhaps all of these combined), and respond in timely fashion to changes that occur in it;
- *pro-activeness*: agents do not simply act in response to their environment, they are able to exhibit goal-directed behaviour by taking the initiative.

This definition is considered as a weak notion of agency. In contrast, a strong notion of agency is one which adds concepts that apply to humans, i.e. mentalistic notions such as knowledge, beliefs and intentions. Autonomy gives an agent the capacity to make independent decisions. This means that each agent possesses its proper temporality in the sense that it runs concurrently to other agents.

Agents must be able to communicate in some way, be it through the environment or directly to one or more agents. This capacity is essential, because agents are always living in an (often structured) environment inhabited by other agents, i.e. in a society of agents that is called a multi-agent system.

3.2.2   Communication between agents

Communication between agents is the capacity to exchange information with other agents. An agent must have a partner to communicate with, either by having an identifier of the partner or by communicating anonymously through predefined interfaces masking the identity of the receiver. The intention or the type of the message is often realised using illocutary speech acts. Examples are request, deny, propose, confirm. It is important to define a common syntax and semantics for expressing the information exchanged. This can be achieved if both agents share a common information model (which is often referred to as ontology or ontological commitments); it is the context of the interpretation of symbols, commonly referred to as an agent communication language (ACL).

The research has led to several agent communication languages whose most popular representatives are KQML (Finin et al. 1994) and FIPA (Foundation of Intelligent Physical Agents) (FIPA, 1997). KQML comes along with KIF (Knowledge Interchange Format) which provides a syntax for message content along with ontologies, FIPA not only includes an ACL but also an agent management system.

For IMPACT (Implementation of Agent for CAC in ATM) we chose the FIPA standard as the basic framework for agent communication.

3.2.3   A FIPA 'flavoured' platform?

IMPACT (Implementation of Agents for CAC in ATM) is concerned with the application of the Intelligent Agents concept as a solution to the Call Admission Control (CAC) problem in Asynchronous Transfer Mode (ATM) networks. The agents developed are lightly based on the Foundation of Intelligent Physical Agents (FIPA) defined concepts, which include the specification of an Agent Management System (AMS), and syntax and semantics of an Agent Communications Language (ACL) amongst other things.

3.2.3.1   Requirements on the agent platform

From the onset an agent platform on which to implement an IMPACT system application was needed. Based on this objective, and given the selection of the FIPA standard as the basic framework for agents, we reasoned that the type of platform that would prove most suitable should have the following features:

- FIPA compatibility;
- be multi-agent and Multi-threaded;
- be able to support other agent architectures;
- a flexible modular design;
- platform source code must be available, for easy extension to the ACL;
- support and detailed documentation.

An evolution path must exist. Working within these requirements, a review of well-known agent platforms (Chauhan, 1997; JATLite, 1997; FIPA'97, 1997) available at that time was conducted, comparing and contrasting their respective functionality accordingly. The results of this extensive review led to the several conclusions (IMPACT Del. 1, 1998):

1. primary amongst those was the virtual absence of any FIPA-compliant platforms,
2. the existence of several layers of complex technology and overhead in some solutions, and
3. the poor definition of the agent communication mechanism, combined with a general inflexibility in others.

3.2.3.2   FIPA mapping

In the light of these revelations, it was decided to investigate a proposal for the development of a new Agent Platform in Java. The underlying rationale for this was having the advantage of controlling the implementation of the agent communication mechanism effectively, with the inherent simplification of the platform, and the subsequent reduction in the technological complexity and list of features. Flexibility in the design of the new platform was essential, however, as it catered for the ability to adapt and blend certain features of the FIPA standard as necessary. This flexible approach to the adaptation of FIPA resulted in the decision to select a subset of the FIPA ACL message performatives, and a sub-set of message parameters. This choice was constrained by the type of system application proposed by IMPACT (IMPACT Del. 2, 1998), and the decision not to allow the Agent Platform to be unduly weighted down with unnecessary performatives whilst still being able to perform some basic FIPA interactions. The nature of the social interactions envisaged communication between agents to include the ability to:

- negotiate with each other,
- invoke actions on each other,
- pass information to each other,
- inform each other if an error occurs.

Consideration of communications in this context implied the careful selection of the sub-set of performatives to facilitate using some FIPA-defined interaction protocols (mainly the FIPA-contract-net, and the FIPA-Iterated-Contract-Net protocol (FIPA, 1997)). This sub-set is presented in Table 3.1.

Table 3.1 FIPA performatives

FIPA Messages Performatives	Main Function
Cfp	Negotiation
Not-understood	Error handling
Propose	Negotiation
Refuse	Action performing
Accept-proposal	Negotiation
Reject-proposal	Negotiation
Request	Action performing
Failure	Error handling
Cancel	Action performing
Inform	Information passing
Confirm	Information passing
Not-understood	Default response

While Table 3.2, below, lists the absolute minimum set of message parameters required for a comprehensible conversation between agents outside of any defined protocols, though the (:protocol) parameter could also be used. We have left out the (:language) parameter, whose main function is to explicitly define the encoding scheme of the (:content) parameter; however, as inter-platform communications perceived from an IMPACT point of view is not implemented, the (:language) parameter value is implicitly assumed, and therefore not specifically referenced.

Table 3.2 FIPA parameters

Message Parameter	Use
:sender	Identify the sender
:receiver	Identify the receiver
:content	Content of the message

Finally, other issues relating to the definition of a FIPA compliant Agent Platform (such as the use of a white and yellow pages service) were examined from an application context point of view. Their introduction into the platform was prioritised according to their immediate relevance. This approach proved to be important, given IMPACT's time constraints and the complexity of the FIPA standard.

3.3   The Java Management Agent: JAM

As already mentioned the Agent Platform is fully implemented in Java. It consists of a graphical user interface, a FIPA component implementing a sub-set of the FIPA performatives, a reactive part to handle FIPA messages asynchronously and a directory facilitator service for white pages service.

Each JAM consists of an applet extending the base class JAMAgent. This base class provides the communication basics (registering the agent, getting the peers) and the common part of the GUI. Additionally each agent inherits from a base FIPA communication class. Each agent has at least one reference to a class that extends the class JAMWorker. The class JAMWorker itself extends the class Thread and implements asynchronous message handling.

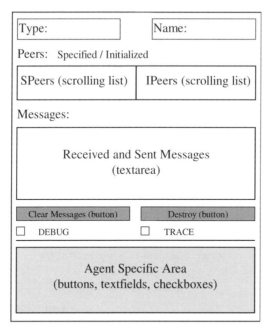

Figure 3.1 Graphical user interface

3.3.1 Graphical user interface

In general, each agent applet has the following outlook (Figure 3.1). On the top the type and the name of the agent are specified. Next there are two lists. The first list names the peers as they are defined in a configuration file. The second list names only those peers that are really available (to which the agent got a reference in the RMI Registry). An important part of the GUI is the message text area, so the GUI was designed to give high priority to it. The messages text area is very useful to observe the interactions. Normally the text is displayed black. In some cases the text changes its colour: a) green text: there was an error during the initialisation of the agent; b) blue text: error during a remote method invocation; c) red text: a request error occurred.

3.3.2 Implementation of the FIPA ACL

To be reasonably FIPA-compliant on the one hand and to make full use of the potential of Java on the other hand, we introduce a novel technique, called signature matching as a basic means for inter-agent communication. In contrast to the FIPA standard, which uses strings as its primary data structure, JAM agents communicate using Java objects. The message content of any JAM FIPA performative is an array of Java objects. Upon reception of the objects the receiving agent determines at runtime using Java's reflection mechanism whether it is capable of handling the received objects. Determining in this context means that the agent verifies whether it has a public method with a signature that matches the types of the received objects. The received objects will be handled by this method. For example an agent may propose (using FIPA's propose performative) to another agent a particular connection request in the form of two Java connection objects C1 and C2. These objects will be handled by an agent's public method with input parameters C1 and C2. Please note that we do not need a return value here, because the result will be sent back in the form of another FIPA performative, e.g. accept-proposal.

The chosen approach has several advantages compared to "classical" RMI-based communication:

- It's close to real-world (agent) communication, because agents propose or inform another agent by simply offering the information without knowing whether the other agent understands it; it's up to the receiving agent to handle the received information accordingly, or to send a not-understood message back, in case it does not know what to do with it;
- It's very flexible, agents do not have to know actual remote method names, but rely on the set of well know FIPA performatives;
- It's less code insensitive, for new or changing functionality, agents do not have to export the appropriate remote functions, but can still use the set of FIPA performatives.

3.4   Asynchronous Message Handling in JAM

There is one thread associated to each instance of an agent class. These so-called worker threads implement the life cycle of the agents. When a remote agent sends a request to an agent peer, then (1) the request is put in the incoming request queue of the accordant 'worker thread' (2a). Therefore the FIPA communication act returns immediately so that the message sender can continue its work (2b). Some time later the 'worker thread' will probably send itself a message to one of its own agent peers (3). This process is shown in Figure 3.2.

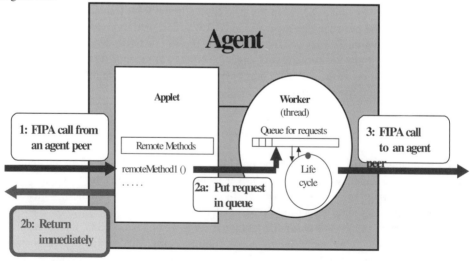

Figure 3.2   Asynchronous Message Handling in JAM

3.4.1   Directory facilitator service in JAM

JAM includes a basic White Pages service realised through a hierarchy of Directory Facilitators used by all agents to get their communication peers. At start-up agents register at their well-known Directory Facilitator Agent and ask for their appropriate communication peers.

3.5   Practical Experiences with Agent Applications

Based on the concept of negotiating and communicating Java agents, we have realized a number of real-world applications. All of these applications could also have been implemented using traditional techniques. However, the use of the ideas and concepts explained in this chapter greatly reduced on the one hand the development time and increased on the other hand the flexibility of the applications in terms of adaptability and extensibility. In the following, two applications are briefly introduced:

3.5.1   Automatic resource brokering and trading service

This service is based on a system consisting of trading agents, customer agents and provider agents, who negotiate and buy on behalf of end-customers quantities of specific resources. Physically, the end-customer agents and the trading agents may be located at the customer location (e.g. home or office), while the provider agents are set up by the providers (telecommunication operators or Internet service providers). A patent is pending on the underlying service and system architecture.

3.5.2   Server-side agents for complex websites

The development of successful e-commerce websites requires that proprietary database systems and complex business logic processes are realised. Clearly, simple shop systems or script-based web solutions (ASP or JSP) are not powerful enough to provide complex web applications. Hence, we applied the concept of communicating Java agents to server-side applications. For a particular project, the resulting architecture was as follows: a Java Servlet running on an Apache webserver environment, provided the front-end of the system. It was the task of this front-end to manage the delivery of HTML pages to the users and tracking of user sessions. The business logic of the service consisted of three communicating Java agents. Each of these agents was given a specific task such as money management or user data management. The use of an agent system made the system extremely flexible in terms of deployment and scalability. It was, for example, possible to apply several load-balancing schemes by varying the number of web servers as well as the number of service agents on the server side.

3.6   Conclusion

In this chapter we presented JAM, a lightweight yet powerful platform for network management applications, which has been developed in the context of the IMPACT project. The platform has already been re-used in other different projects and has therefore proven its flexibility.

However, the current implementation can be seen as a solid starting point for a true agent platform. For future work we therefore envisage a full implementation of the FIPA ACL which was not possible due to the time constraints, the main focus of the IMPACT project and the time required to integrate the FIPA Agent Management System.

Additionally we envisage transforming the JAM platform into a Java Bean-compliant component framework, and updating it to reflect the recent changes in the FIPA standard.

3.7   References

Chauhan, D. (1997) JAFMAS: A JAVA-based Agent Framework for Multiagent System Development and Implementation, *PhD Thesis*, ECECS Department, University of Cincinnati, (Available at http://www.ececs.uc.edu/~abaker/JAFMAS).

Finin, T., Fritzon, R., McKay, D. and McEntire, R. (1994) KQML as an Agent Communication Language. In *Proceedings of the 3rd International Conference on Information and Knowledge Management (CIKM)*. ACM Press.

FIPA '97 (1997) Foundation of Intelligent Physical Agents, http://www.fipa.org

IMPACT Deliverable 1 (1998) Specification of Agents, AO324/QMW/IMP/DS/P/003/B2. Available at http://www.acts-impact.org/

IMPACT Deliverable 2 (1998) Specification of Scenarios, A0324/QMW/IMP/DS/P/011/B2, 1998. Available at http://www.acts-impact.org/

JATLite, (Intelligent Agent toolkit) http://JAVA.Stanford.EDU/JAVA_agent/html/

Wooldridge, M. and Jennings, N. R. (1995) Intelligent Agents: Theory and Practice. *The Knowledge Engineering Review*, Vol. **10(2)**, 115-152.

4

Adaptive QoS Management via Multiple Collaborative Agents

A. Bordetsky

4.1   Introduction

Current advances in such technologies as "wearable augmented reality systems" (Lawlor, 1999) demonstrate emerging opportunities in bringing multipoint collaborative voice, video and data communications into the environment of mobile wireless networking. In such environments mobile users become independent information nodes of a rapidly changing wireless network and presence of an augmented reality system makes multimedia communication feasible even through low-bandwidth wireless links.

In order for such multipoint multimedia wireless services to effectively evolve, it becomes increasingly important to use service management tools based on the multiple cooperating agents (Bordetsky and Lewis, 1999) that can support Quality of Service (QoS) adaptation in the highly dynamic environment of networking resources and customer application profiles. For the software agents representing multiple independent nodes of mobile wireless network this would include response time management, rapid re-configuration, and in some cases (e.g. IP over ATM) dynamic bandwidth allocation in accordance with content and customer communication profiles.

In this chapter we discuss the models for multiple agent individual adaptation and their adaptive cooperative behaviour. The proposed models are based on the concept of integrating Service Management Layer and Networking Layer feedback (Bordetsky, Brown and Christianson, 2000). Case-based reasoning memory is used to capture and integrate the feedback controls for different TMN layers that agents could share in association with customer profiles and QoS solutions.

Multiparticipant group-decision support models (Marakas, 1998) are used to structure the agents' cooperative behaviour in the form of the agent committees. The QoS committee solutions are captured into the artificial neural network model that is integrated with the case-memory feedback controls stack.

We will start by discussing the feedback control solutions for the individual agent adaptation in the multiple agent architecture.

4.2   Layers of Feedback Control: Individual Agent Adaptation

We can address the problem of agents' adaptation by tying the TMN (Aidarous and Plevyak, 1995) Service Level Management functionality (see Figure 4.1) to the fundamental concept of system coordination, which identifies critical relationships (Malone and Crowston, 1994) by revealing associated feedback controls. From this perspective the process of adaptive control and coordination in the multiple agent architecture could be based on the idea of mapping feedback control relationships into an agent's shared awareness memory, and delivering feedback controls via an ensemble of agent facilitators. In structuring the agents as agent facilitators with bridging, routing and gateway functionality we follow the evolving KQML concept (Mayfield, Labrou and Finin, 1995) of agent communication models (Genesereth and Ketchpel, 1994). In the models below we expand the bridging, routing and gateway functionality into the agents' integration with case memory. Case memory supports the learning of feedback control relationships and adaptive management of QoS requirements by utilising a case-based reasoning technique (Lewis, 1995; Bordetsky and Bourakov, 1998) for indexing, capturing and retrieving the feedback structures associated with Web conferencing events and QoS constraints.

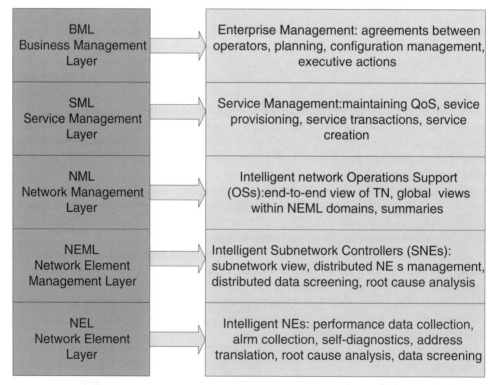

Figure 4.1 TMN intelligent management architecture

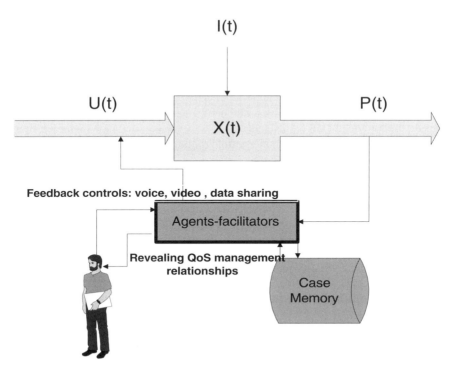

Figure 4.2 Feedback control model for individual agent adaptation

Real-time applications such as audio and video conferencing and shared application control, however, have stringent requirements regarding maximum delay and minimum bandwidth. Applications that are asynchronous in nature can adapt naturally, leading only to changes in response time. Real-time applications, however, may choose to reduce the quality of the data stream to reduce bandwidth needs.

When we consider multiple applications running simultaneously, lower-priority applications may be required to adapt to lower-bandwidth usage or be switched off entirely to free up bandwidth for higher-priority applications. We respectively consider two layers of feedback controls: *Call Preparation Control* (CPC) and *Connection Control* (CC). Call Preparation Control integrates feedback gathered from previous conferencing sessions to make informed decisions regarding connection setup and bandwidth tradeoffs in future sessions. Connection Control reflects ongoing performance measurement and adaptation throughout the length of the call.

Call Preparation Control requirements to support multimedia multipoint applications include:

- A call will have to establish, modify, execute and terminate voice, video and application-sharing communication between multiple users.
- A call involves coordination between parties to satisfy their response time, bandwidth, and other QoS requirements.

- A call contains relationships between user profiles, media and system resources. These relationships may be dynamically modified during a call.
- Each user can request resources individually.
- A call will allow negotiations between different sites for system resources.

Connection Control requirements could be summarised as follows:

- Supervising provided QoS parameters.
- Providing flow control, congestion control, routing, reservation and renegotiation of resources.
- Modifying and releasing connections.

In terms of the length of change effects, Call Preparation Control adaptation could be referred to as long-term adaptation, mainly associated with allocating resources for the entire length of a multimedia call. Conversely, Connection Control adaptation would deal with short-term adaptation, which might be required many times during a single call.

4.3   Call Preparation Adaptation: Service-Layer Feedback Controls

The architecture of the proposed adaptive management mechanism is represented by three components: a case-based reasoning memory, agent facilitators and collaborative feedback controls (see Figure 4.2). The layers of case memory are structured according to the feedback control relationship for a Web conferencing service:

$$SLM\_Event(t)=\{U(t), X(t), P(t), I(t)\}, \tag{4.1}$$

where:

$SLM\_Event(t)$ stands for a Service Level Management event, $X(t)$ is a set of SLM process state variables (QoS constraints such as response time and bandwidth), $U(t)$ is a set of user input controls (e.g. desktop video conferencing calls, links to knowledge sources), $P(t)$ is a set of service process outputs (e.g. the content of an electronic commerce transaction), $I(t)$ describes the environmental impact to the service management process.

In accordance with the layered memory architecture of agent facilitators, agents are divided into bridge or router agents which operate with different combinations of feedback control layers. A Bridge Agent typically provides multicasting of $P(t)$ content and/or $X(t)$ information only, whereas a Router Agent associates the Web conferencing feedback controls with output/state memory frames content:

$$\{U(t), User\ View\ (SLM\_Event(t))\} \tag{4.2}$$

The Router Agent plays a major role in providing feedback controls and adaptation in service management. It provides user-memory transactions, and supports capturing of communication parameters, personal, document and task profiles. It enables location of appropriate human sources of knowledge and manages desktop video conferencing calls to selected experts. It provides training and capturing of QoS management knowledge in case memory. Figure 4.3 illustrates capturing feedback control association of P(t) and X(t) with U(t) into the case memory.

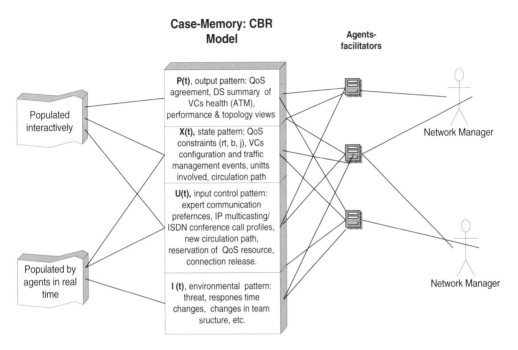

Figure 4.3   Feedback control association {P(t), X(t), U(t)} and QoS constraints in case-based reasoning memory.

The knowledge retrieval model is a hierarchy of case memory layers (Figure 4.3), in which each interface between layers (from the bottom up) is an association based on the underlined feedback structure. The content profiles and user response time requirements are captured in real time and populate the lower segment of the case memory stack. Sequence of application calls (content profile) and time stamps captured by an agent (see Figure 4.4) are converted into response time and bandwidth requirements that populate the QoS segment of a case memory frame.

Suppose that the QoS constraints associated with the specific SLM_Event comprise the boundaries that define preferred bandwidth for voice, boundaries that define preferred bandwidth for video, boundaries that define preferred bandwidth for white board, and boundaries that define preferred bandwidth for the application sharing.

According to such a profile, each conferencing node has associated voice, video, white-board and application-sharing delivery trees. Switching between these delivery trees could help to satisfy otherwise infeasible response time requirements.

Suppose that rules for switching the streams are identified based on the operations heuristics, or otherwise quantitative experiments (Bordetsky and Bourakov, 1998). They correspondingly are populated into the SLM_Event associated QoS segment of the agent's case memory.

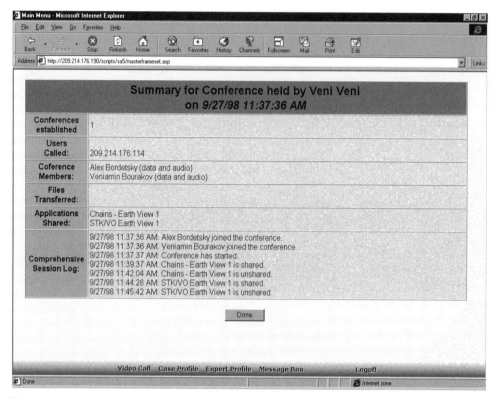

Figure 4.4   Agent-router: mapping X(t) to U(t) by capturing the response time requirements and content profiles into the agent's case memory

The Router Agent reads the QoS segment of the feedback control association {P(t), X(t), U(t)} from the case memory (Figure 4.3) and needs to coordinate the delivery tree switching (i.e. bandwidth allocation solutions) with the other agent facilitators. When coordination is done (this model is described in section 4.8) the agents transfer the coordinated solution to the network-layer connection control.

4.4   Connection Control Adaptation: NE Layer Adaptation

In general the Connection Control requirements include:

- Supervising QoS parameters,
- Providing flow control, congestion control, routing, reservation and renegotiation of services,
- Modifying and releasing connections, and
- Notifying applications to allow them to adapt.

As opposed to the Call Preparation Control, in which decisions are made *before* the call is made, Connection Control is done on an ongoing basis throughout the duration of the call. Feedback regarding network conditions must be continuously collected and processed in order to allow the applications in use to adapt. The most dynamic network resource in wired and wireless networks is allocated channel bandwidth. This is where we concentrate our efforts in network-layer feedback controls.

Brown and Christianson in Bordetsky, Brown and Christianson (2000) provide a detailed review of different network-layer control models. The review includes Consenting Equal Division (Parris, Ventry and Zhang 1993), Hierarchically Layered Data Stream (Hoffman and Speer, 1996), Self Organised Transcoding (Kouvelas, Hardman and Crowcroft, 1998), among others.

By looking at the deficiencies of these tools, such as: not addressing multicast (Parris, Ventry, and Zhang, 1993), not allowing learning from the previous connection adaptation Stream (Hoffman and Speer, 1996), or requiring similar bandwidth allocation for collocated receivers (Kouvelas, Hardman and Crowcroft, 1998) the authors suggest hierarchical encoding as a perspective model for network and service layer's integrated adaptation.

The results of the study described in Bordetsky, Brown and Christianson (2000) demonstrate that bandwidth adaptation can be used to match offered load to allocated bandwidth. It is reasonable to expect that, over time, many calls will exhibit similar behaviour. The current adaptation mechanism does not learn from experience, either from events which take place over the course of the current call or from previous calls. It is here that Call Preparation Control methods may be used most effectively.

4.5   Integration of Service Management and Network Management Layer Adaptation

Using the example from the previous section, we can identify several scenarios where Service Management Layer (Call Preparation Control) adaptation output would be useful:

*   Specifying the initial number of multicast groups to which to subscribe.
*   Specifying the number of consecutive report intervals which should trigger adaptation.
*   Specifying the levels of loss which are significant, for indicating both congestion in the network and the absence of congestion.

As most multipoint conferences will consist of many components including audio, video and shared-application control, it will also be necessary to balance the bandwidth needs of each individual tool. In this way, video streams may be constrained to black and white images in favour of high-quality audio or lower-priority streams may be shut off entirely in favour of higher-priority streams.

Let each loss ratio report for four multimedia components, audio, video, shared application and whiteboard, be represented by the vector:

$$\Delta P_i = \left(\Delta p_{i1}, \Delta p_{i2}, ..., \Delta p_{in} \right), \tag{4.3}$$

where $i = 1, 2, 3, 4$; Δp_{i1} denotes the loss ratio, Δp_{i2} stands for the highest sequence number received, and Δp_{i2} could be used to specify jitter.

The loss ratio, highest sequence number received, and jitter values calculated for each RTP receiver report are logged and made available for post-processing.

If the observed combination of $\{\Delta p_{ik}\}_i$ values is judged acceptable for processing without immediate bandwidth adjustment, then the agent-facilitator sets up inequality as negative. If an expert (e.g. a network manager/operator) evaluates this vector as indicating that bandwidth adaptation is necessary, then a non-negative value is set up. The expert responses consolidated during the knowledge acquisition (training) phase would constitute an integrated system of the form:

$$\sum_{j=1}^{n} (W_{ij} \times \Delta p_{ij}) \geq 0$$

$$\sum_{j=1}^{n} (W_{ij} \times \Delta p_{ij}) < 0 \tag{4.4}$$

Solution vector $W = \{W_{ij}\}$ for system (4.4) is used to identify the filter as a discriminant linear function for audio, video, shared application and white board streams:

$$W_i \times \Delta P_i \geq 0 \tag{4.5}$$

In many cases, the same training vector ΔP_i could be evaluated as satisfactory for the video stream (i.e. no need to initiate short-term bandwidth adaptation), but at the same time be evaluated as requiring bandwidth adjustment for the voice stream. This would create conflicting constraints in system (4.5) and would result in a state of unfeasibility. When system (4.4) becomes infeasible, it is not possible to identify a single discriminant function (4.5). The solution requires a *set* of QoS discriminant functions that represent multiple participating agents.

4.6   Adopting the Group Decision Support Techniques: A Committee Model

In practice the collaborative multiple agent architecture will be used in conjunction with network operations management teams' decision support relationships. It seems natural, therefore, to look at the perspective collaborative multiple agent structures using the multiparticipant information processing and networking paradigm. This approach was originated in early works of Galbraith (1974), Tushman and Nadler (1976) and later implemented by many researchers. Provided that in an organisational setting the multiparticipant (Holsapple, 1991) decision-making relationships could take place locally or span across vertical and horizontal organisational boundaries, it seems natural to attract known topologies of local-area networking and wide-area networking, as well as routing, switching and multicasting metaphors to describe collaborative multiparticipant relationships.

Based on the study of two dimensions, the direction of information stream and the structure of information flows in group decision making. Marakas (1998) suggests the hierarchical classification of multiparticipant decision-making structures. In this classification different multiparticipant decision-making dependencies comprise three basic models: group, team and committee.

In the group model (Figure 4.5) the structure of information flows is a mesh network. It links multiple decision-makers in a way that allows complete interaction among them.

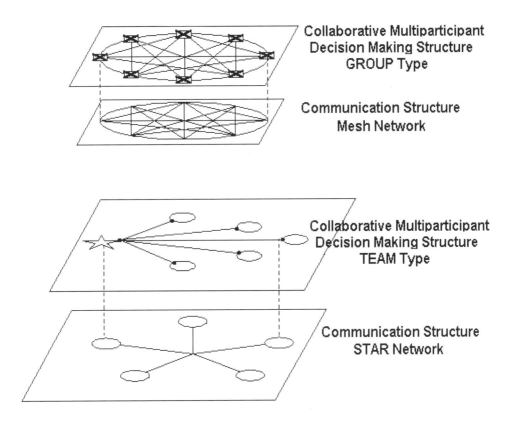

Figure 4.5 GROUP-type and TEAM-type multiparticipant decision making structures

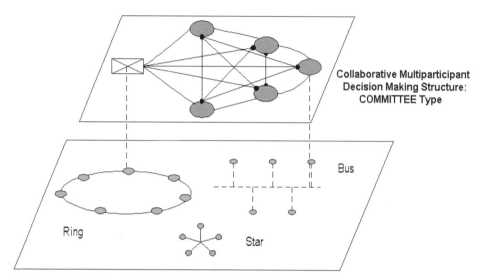

Figure 4.6 COMMITTEE Structure

The third basic model is committee (Figure 4.6). It combines a single decision-maker with complete participant interaction. It allows collective behaviour that is based on different types of majority rules or consensus protocol. A combination of star and ring topologies could be used to support local and interdepartmental committee structures. The group structure could be viewed as comprising several committees.

Group multiparticipant structure may not be the most appropriate prototype for the multiple agent adaptation as it relies on the mesh topology and doesn't separate the facilitator (coordinator) from other members. Unlike it, team topology naturally allocates a role for the decision-maker (facilitator), but it lacks cooperative relationships among the members, which could be critical in the joint knowledge discovery process. From that stand-point the committee model represents a reasonable compromise between the group and team multiparticipant structures. It allows a facilitator (coordinator) role and compensates for the lack of participants' interaction that is typical for the team structure.

Based on the described consideration we select a committee model for structuring the agents' collaborative process.

4.7   Modelling the Agents' Committee

Let combined QoS constraints that different agents are responsible for be represented in the following generalised form:

$$M_k = \{x \in R^n \mid RL_k \leq c^k x \leq Al_k, \ x \geq 0, \ [RL_k, AL_k] \in R^m\}, \ k=1,..., p; \qquad (4.6)$$

where,

$$M = \varnothing, \; M = \bigcap_{k=1}^{p} M_k,$$

and each agent r, $r = 1, ..., s$, operates the subset of bounds $\{[RL_{rk}, AL_{rk}]\}$, such that, different agents can share the same constraints:

$$(\bigcap_r \{[RL_{rk}, AL_{rk}]\}) \neq \varnothing.$$

There are two known models of isolating a portion of model (4.6) that would allow for the committee to compromise on changing the boundaries for conflicting constraints (discriminant functions for the agents): minimal infeasible (Chinneck and Dravniecks, 1991) and maximal feasible subsystems (Bordetsky, 1996; Mazurov, 1991).

Maximal under inclusion feasible systems of discriminant functions could be especially useful for representing the committees of agents. According to (Mazurov, 1991) a committee of solutions is a finite or infinite set of elements, such that each constraint is satisfied by a majority of its members.

Consider the inclusions:

$$x \in D_j \, (j \in J), x \in X \tag{4.7}$$

If $0 < p < 1$

$$|K \cap D_j| > p|K| (j \in J) \tag{4.8}$$

then $|K|$ is called a p-committee of system (4.7). One of the most practical representations of the p-committee is a selected set of solutions of its maximal feasible subsystems.

The committee of constraints is a selected maximal feasible subsystem,

$$I_k = \{i \in I_q \subseteq I \mid \underset{\{I_q\}}{Max} \sum_{i \in I_q} C_i \} \tag{4.9}$$

$$I_q = \{i \mid \bigcap_{i \in I_q} M_i \neq \varnothing \; \& \; \bigcap_{i \in I_q} M_i = \varnothing \}, I'_q = I_q \bigcup (i \in I\backslash I_q),$$

where C_i, is a constraint (or agent) 'weight' designating its relative importance.

Based on the described committee model we can form an artificial neural network that would provide a hierarchical structure of discriminant functions capable of learning changes in the W_i coefficients for the system of QoS constraints (4.6). The constraints are managed by cooperating agents for allocating resources to audio, video, shared-application and white board streams.

4.8   Learning Agents' Collaborative Experience by ANN

How can we facilitate learning and upgrading of W_i solutions for the set of QoS discriminant functions (4.4)? We implement the following model of a four-layer Artificial Neural Network (ANN) that provides a hierarchical structure of discriminant functions capable of learning changes in the W_i coefficients (Figure 4.7).

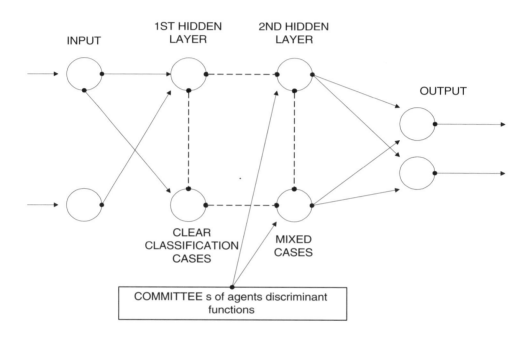

Figure 4.7 Representing the agents' committees by Artificial Neural Network

4.8.1   Input layer

The *input layer* represents the learning vector ΔP_i in which each input node stands for an aspiration reservation interval for a single constraint $[RL_k, AL_k]= \Delta p_k$ (e.g. loss ratio interval, jitter interval, etc.)

4.8.2   First hidden layer

The first hidden layer represents the discriminant functions for the revisions $\{\Delta p_k\}$ that experts evaluate as 'good' or 'bad' for initiating network-layer bandwidth adaptation without any contradiction. Each of the nodes in the *first hidden layer* represents *one* linear discriminant function W_i x $\Delta P_i \geq 0$ that exactly separates 'good' and 'bad' revisions of $\{[RL_k, AL_k]\}$ intervals. Weights w_{ij} which are the coefficients of discriminant functions, are subject to changes in the process of training and are determined as feasible solutions for a system of constraints in a training sequence (4.4).

4.8.3   Second hidden layer

Nodes of the *second hidden layer* match the training cases in which revisions of $\{[RL_k, AL_k]\}$ intervals for the shared constraints are conflicting, e.g. patterns of 'good' and 'bad' QoS are overlapping. In this case, the set of training constraints is infeasible. Each of the nodes in the second hidden layer represents a committee of discriminant functions. This is a committee of solutions, where the set of weight vectors satisfies more than half of the inconsistent constraints in the system.   More precisely, each node of the second hidden layer has a threshold function:

$$F(\underline{w}) = \Sigma_k \, sign \, (\underline{W}_{k,*} \, \Delta \underline{P}) \tag{4.10}$$

where sign(.) = $\{1, 0\}$. If $F(\underline{w})$ > (m+1).r, where **m** is the number of members in the committee $\underline{w} = [\underline{w}_1, ..., \underline{w}_k, ..., \underline{w}_p]$, and r is the ratio of participation (usually one half). When the node fires, the adjacent vectors $\underline{w}_i$ are taken as the coefficient vectors for related constraints.

The selection criteria for the committee of constraints may vary. In the case where weights are equal, the selection criterion is a simple majority rule. The learning process will produce the union of the initial discriminant functions and the set of developed (learned) empirical constraints that represents network-layer bandwidth adaptation experience. By capturing and updating such constraints concurrently with Connection Control sessions, the neural net will represent an adaptive filter for interfacing short-term feedback controls with Call Preparation controls captured into the case memory.

4.9   ANN Integration with CBR Memory

For a single agent the discriminant function (4.5) is placed into the QoS segment of the case-based memory (see Figure 4.3) stack that contains the associated segment of application-layer feedback controls and user profiles.

When the Connection Control process begins, agent-facilitators check the observed values of ΔP by plugging them into the discriminant function (4.5). If the value of W_i x ΔP_i is positive, the agent-facilitator transfers control to the network-layer bandwidth adaptation tool for providing immediate bandwidth adjustments.

When the Connection Control adaptation starts, an agent-facilitator checks the observed values of ΔP by plugging them into the first-layer discriminant functions (4.5). If, for all nodes, the value of W_i x ΔP_i is positive, the agent-facilitator transfers control to the network-layer bandwidth adaptation tool for providing immediate bandwidth adjustments. If some nodes vote 'yes' to bandwidth adjustment, and the others vote 'no', then the second-layer agent committee nodes that indicate associations with the current multimedia call profile are checked. If the agents' committee node votes 'yes', then the network-layer bandwidth adaptation tool is turned 'on'.

4.10   Future Directions: Orbital Agents

There are many ways of applying the described collaborative agents architecture to the emerging mobile wireless communications. Initial development of the described models started on the basis of the TELCOT Institute testbed (Figure 4.8) with access to the National Transparent Optical Network and the NASA Advanced Communication Technology Satellite (Figure 4.9).

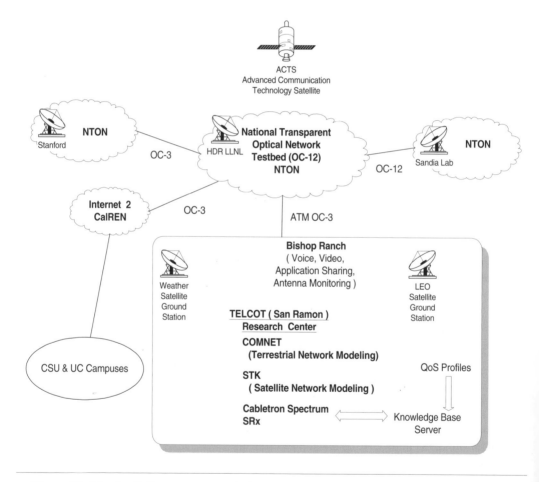

Figure 4.8 Adaptive QoS management testbed

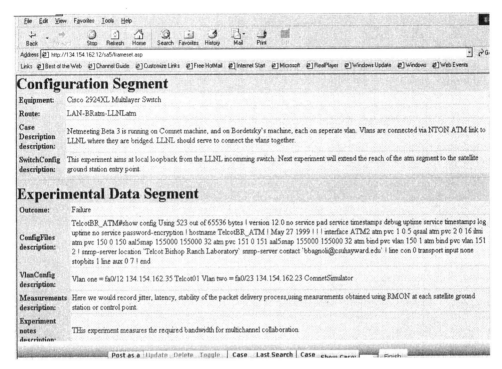

Figure 4.9 Agent-facilitator U(t) recommendations for configuring an ATM satellite based on the TELCOT's testbed experiences

The agents were allocated on the terrestrial Network Operations Centre (NOC) site with the simulated Mobile Switching Centre (MSC) functionality (Figure 4.10). It seems natural to look at more advanced management environments in which cooperating agents could be downloaded to the mobile terminals, or they could be distributed across the multiple MSCs. They could be portable and could move between the mobile terminals as well.

At the extreme of such advanced management solutions is the orbital implementation of the discussed adaptive management models. This project is under way in cooperation with Stanford Space Systems Development Lab. It is based on unique management nodes carrying on the agent's formation of the picosatellites.

At present, in order to have private multipoint networking, especially if it includes collaborative multimedia applications, one must use primarily terrestrial multipoint services based on the terrestrial MSCs. Terrestrial links to the IP multicasting routers, MCUs and ATM switches limit the capability of network management infrastructure to respond to the threat by restructuring direct access of geographically distributed collaborating teams to the routers, MCUs and switches. Terrestrial agent-hosts, MCUs, Mbone routers, and ATM switches are not mobile. However, small low-orbit single-function hosts could provide an unparalleled opportunity for geographically dispersed groups to configure their private multipoint collaborative environment in a hostile environment. For the small number of mobile geographically distributed users who are in need of conducting multipoint real-time audio/video and data-sharing ISDN/IP conferences, the closest MCU site could be in orbit. The fleet of 3-9 picosatellites presents an attractive innovative orbital solution to this problem. The constellation of picosatellites is assumed to exhibit cooperative behaviour by integrating simple functions of individual small satellites into the complex multiple satellite agent's performance.

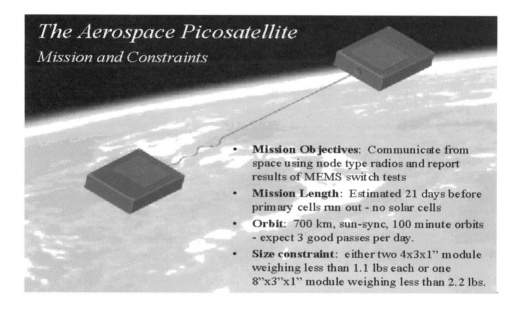

Figure 4.10 First picosatellites

Such a constellation of several picosatellites would be an ideal platform for allocating intelligent agents and inexpensive switching hosts into the orbit. Figure 4.10 illustrates the first picosatellites recently tested by the Aerospace Corp. The author had the good fortune to participate in the first operations management experiments with these satellites.

In such a single function satellite formation one picosatellite would be an Agent-router capable of configuring the multipoint collaborative session based on the user and task profiles, coverage pattern, state of the traffic, and immediately available orbital hosts. It will communicate with a MCU picosatellite for multipoint circuit switching session or with a Mbone router picosatellite for an IP multicasting multipoint call. It will contact the Probe picosatellite for coverage pattern and traffic analysis in the area. It might need to communicate with a Codec picosatellite for compressing the image for downlink in a way similar to an external modem that is used with several computers. The Agent satellite would populate the acquired data into the constellation knowledge base that is carried on by the Memory picosatellite. It would store the adopted multipoint configuration solution in the memory to share the solution with other orbital Agents, or to keep it until the next session as a prototype. In the highly dynamic environment of urban warfare immediate access to the previously found configuration solutions (which could be quickly adopted subject to current session task, coverage and traffic patterns) could dramatically improve the responsiveness in configuring and managing multipoint collaborative environments.

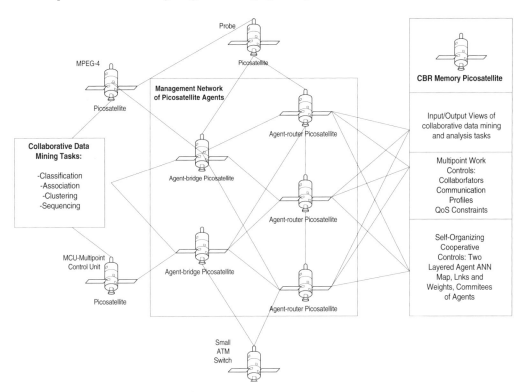

Figure 4.11 Picosatellite-based orbital agents

We envisage that several Agent pico satellites will operate in the proposed intelligent orbital management system. They could be tight to the specific information analysis/ intelligence task, or could be multiple copies of the same basic Agent picosatellite. They could be approached simultaneously (or with minimal latency requirements) by different dispersed groups for establishing a multipoint collaborative session. In such a case they would start sharing solutions retrieved from the constellation memory and would need to behave cooperatively. The cooperative behaviour of intelligent orbital agents will be modelled by unique orbital Artificial Neural Network (ANN) architecture in which each Agent picosatellite will have a functionality of an ANN node. A Memory picosatellite will keep ANN links and their weights that Agent picosatellite will be upgrading as a result of their cooperative learning (Figure 4.11).

4.11   References

Aidarous, S. and Plevyak, T. (1995) *Telecommunication Network Management into the 21st Century*, IEEE Press, New York.

Bordetsky, A. B. (1996) Reasoning on Infeasibility in Distributed Collaborative Computing Environment, in H. Greenberg, Ed., *Annals of Mathematics and Artificial Intelligence*, **17**, 155-176.

Bordetsky, A. and Bourakov, E. (1998) Agents-Facilitators for Adaptive Management of Collaborative Environments, in *Proceedings of the 3rd INFORMS Conference on Information Systems and Technology,* April 26 28, Montreal, pp. 82-96.

Bordetsky, A. and Lewis, L. (1999) Knowledge-Based Support for Collaborative Management of High-Speed Telecommunication Networks, in *Proceedings of the 4th INFORMS Conference on Information Systems and Technology,* Cincinnati, OH, pp. 119-130.

Bordetsky, A., Brown, K. and Christianson, L. (2000) Adaptive Management of QoS Requirements for Wireless Multimedia Communications, in *Proceedings of 8th International Conference in Telecommunication Systems*, Nashville, TN.

Chinneck, J. W. and Dravniecks, E. W. (1991) Locating Minimal Infeasible Constraint Sets in Linear Programs, *ORSA Journal of Computing,* **3(2)**, 157-168.

Galbraith, J. (1974) Organisation Design: An Information Processing View. *Interfaces,* **4,** 28-36.

Genesereth, M. R. and Ketchpel, S. (1994) Software Agents, *Communications of the ACM,* **37(7)**.

Hoffman, D. and Speer, M. (1996) Hierarchical Video Distribution over Internet-style Networks, in *Proceedings of the IEEE International Conference on Image Processing,* Lausanne, Switzerland, Sept.

Holsapple, C. (1991) Decision Support in Multiparticipant Decision Making. *Journal of Computer Information Systems,* Summer, 37-45.

Kouvelas, I., Hardman, V., and Crowcroft, J. (1998) Network Adaptive Continuous-Media Applications through Self Organised Transcoding, in *Proceedings of Network and Operating Systems Support for Digital Audio and Video (NOSSDAV 98),* Cambridge, UK, 8-10 July.

Lawlor, M. (1999) Today's Exploration Launches 2001: A Battlespace Odyssey, *SIGNAL,* December, pp.17-20.

Lewis, L. (1995) *Managing Computer Networks: A Case-Based Reasoning Approach*, Artech House.

Malone, T. and Crowston, K. (1994) The Interdisciplinary Study of Coordination, *ACM Computing Surveys*, **6(1)**, 87-119.

Marakas, G. (1998) *Decision Support Systems in the Twenty-First Century*, Prentice Hall.

Masurov, V. M. (1992) Decomposition in Committee Pattern Recognition Constructions, News of USSR Academy of Science. *Journal of Computer and Systems Science,* **4,** 162-170.

Mayfield, J., Labrou, Y. and Finin, T. (1995) Desiderata for Agent Communication Languages, in *Proceedings of the AAAI Symposium on Information Gathering from Heterogeneous, Distributed Environments, AAAI-95 Spring Symposium,* Stanford University, Stanford, CA March 27-29.

Parris, C., Ventre, G. and Zhang, H. (1993) Graceful Adaptation of Guaranteed Performance Service Connections, in *Proceedings of IEEE GLOBECOM '93*, Houston, TX, November.

Tushman, M. and Nadler, D. (1977) Information Processing as an Integrating Concept in Organizational Design, *Academy of Management Review,* July, pp. 613-624.

5

Agent-Mediated Trading: Intelligent Agents and E-Business

M. Klusch

5.1   Introduction

The electronic marketplace provided by the Internet and Web is about to establish itself as a significant economic factor worldwide. It does not only leverage relationships for strategic advantage in global commerce, as witnessed by increasing mergers and acquisitions of new information technology and Internet-based companies but, more strikingly, it also enables the formation of virtual enterprises. These virtual enterprises are capable of producing and delivering value-added services and products tailored to individual just-in-time customer preferences as well as building virtual auction houses for both traditional and reverse auctioning for the common user. On the other hand, it raises customer expectations regarding issues of new quality of shopping experience and product and service quality of new brands on the Internet. Such brands can be destroyed as quickly as they have been established. This demands for advanced solutions, not only for the rather traditional business-to-business/customer electronic commerce, but also for integrated commerce in the future. The latter means the flexible and efficient coordination of operational business processes throughout the entire supply chain, encompassing customer product online orders and manufacturing of the desired product as well as its shipment to the customer.

Intelligent information agents (Klusch, 2000a) are autonomous computational software entities which have access to multiple, heterogeneous data and information sources, and pro-actively acquire, mediate, and maintain relevant information on behalf of their users or other agents. Information agents are especially meant (1) to provide a pro-active resource discovery, (2) to resolve information impedance of information consumers and providers, and (3) to offer value-added information services and products. That includes, in particular, retrieving, purchasing, filtering, fusing, and presenting relevant information to human decision-makers on demand and preferably just in time. The situation becomes even more complex since customers as well as merchants' products, services, and quality may change rapidly over time.

Though e-business, including e-commerce on the Web, is not the original, classical application domain of information agents, it certainly is the most steadily growing one. However, e-business might happen without any intelligent agents if agent technology in general fails to be injected into currently emerging Internet-mediated transaction standards and systems. That is independent of the fact that the integrated usage of personalised digital assistants in future e-business applications is expected to be more convincing, more attractive, and more qualitative for the common users' everyday business on the Web.

In the remainder of the chapter, section 5.2 provides the latest facts and figures on the domain of e-commerce. In section 5.3 we give a brief introduction to intelligent information agents, and then focus in section 5.4 on selected enabling techniques and examples for agent-mediated trading before concluding the survey in section 5.5.

5.2   E-Commerce: Some Facts and Figures

Electronic commerce (Merz, 1999) may be defined as the set of activities of trading goods and services online. It can be structured into the market segments: business-to-customer (B2C) such as online shops and auctions provided by portals, business-to-business (B2B), customer-to-business (C2B) such as reverse auctions, and customer-to-customer (C2C) e-commerce. As such it is part of electronic business which covers a broader range of issues including any business process and transaction on the Internet devoted to, for example, call centres, corporate intranet, customer relationship, and supply chain management. E-commerce has steadily grown since 1995. According to recent market research reports of Meta Group and Forrester Research, between 200 and 350 billion US$ e-commerce sales are expected in 2000 in Europe and the States, respectively. Remarkably, the B2B market segment of e-commerce is predicted to still outweigh that of B2C worldwide by a magnitude, with in the region of 1.3 and 0.1 trillion US$, respectively. On the other hand, the consumer online spending rose from 7 billion US$ in the whole holiday season of 1999 up to 2.8 billion US$ in the month of January 2000. In any case, up to 71% of companies worldwide are expected to link to the electronic marketplace on the Internet, generating up to 20% e-business-based turnover on average by 2004. Of course, these numbers are due to economic shift changes but should clearly illustrate the long-term potential of e-business and e-commerce.

Basic key enabling technologies for the development of any e-business solution concern:

*   *standard data representation, retrieval and exchange*, like XML, UML, RDF(S), EDIFACT/WebEDI/EDIINT, standardised domain ontologies across domain boundaries (Gruber, 1992; Steels, 1998), methods for automated semantic mapping of heterogeneous ontological knowledge minimising loss of information (Sheth et al., 1996), data retrieval, and data mining tools,
*   *secure user profiling and data*, like OPS (open profiling standard), W3C's P3P (platform for privacy preferences), (a)symmetric coding schemes, digital signatures, and digital watermarks,

- *secure electronic payment*, like VISA/MC's SET for payment with credit card, digital cash like DigiCash's eCash, DEC's MilliCent, and smart cards, or deduction from a given customer account like at central virtual markets or cyber-malls, and
- *standard protocols covering most issues of electronic trading*, like IETF's IOTP (Internet open trading protocol), OTP (open trading protocol), and OBI (open buying on the Internet).

Any emerging consensus on an *accounting and pricing* structure, such as flat rate, capacity-based or usage-sensitive pricing, is as important as effective trust and security mechanisms to facilitate e-commerce transactions in a digital economy. Another challenge is how to model, measure, and reason on *trust* in e-commerce settings (Manchala, 2000). Notably, according to opinion polls conducted by Odyssey and Forrester Research Inc. in 1999, about 85% of all customers evaluated their online buying experience as 'very good', in particular trusting and showing more loyalty to online than traditional offline retailers.

Different trading models and schemes may be compared along (1) the design on economic principles such as dominant, competitive and adaptive strategies and equilibria, (2) privacy of interests of the participants and anonymity of identities, (3) complexity of trading mechanisms in terms of computation and communication. In order to apprehend e-commerce and business in a more homogeneous way a variety of different architectures, frameworks, and reference models have been developed and are under ongoing research. These include, for example, the CBB (consumer-buying-behaviour) model developed at MIT (Guttman, Moukas and Maes, 1999), CommerceNet's eCo framework,[1] OMG's BOCA (business object component architecture),[2] and the SEMPER (secure electronic marketplace for Europe) layered architecture comprising services for trading at different stages and levels of complexity[3] However, each of these approaches appears to be either too generally specified or too specific in part; an efficient, dynamic refinement of interfaces and service components by the user is, if at all, scarcely supported.

5.3   Intelligent Information Agents in Brief

Information agents are a special kind of so-called intelligent software agents. Software agent technology that originated from distributed artificial intelligence is inherently interdisciplinary. Thus, the notion of agency is quite broadly used in the literature; it might rather be seen as a tool for analysing systems, not an absolute characterisation that divides the world into agents and non-agents. However, *intelligent agents* are commonly assumed to exhibit autonomous behaviour determined by its:

- *pro-activeness*, which means taking the initiative to satisfy given design objectives and exhibit goal-directed behaviour,
- *reactive or deliberative actions*, which means perceiving the environment and using timely change management to meet given design objectives, and

---

[1] http://eco.commerce.net/

[2] http://www.d-a-t.com/Download/standards%20info/standards_and_specifications_dow.htm

[3] http://www.semper.org/

- *social cooperation* in groups with other agents and/or human users when needed.

It depends on the concrete application domain and on the potential for a solution for a particular problem what an intelligent agent in practice is supposed to do. Today, intelligent agents are deployed in different settings, such as industrial control, Internet searching, personal assistance, network management, games, software distribution, and many others. For a more comprehensive and introductory literature on intelligent agents we refer to (Wooldridge, 1999; Jennings and Wooldridge, 1998).

Agent technology is quite on its way to producing mature standards concerning software agent architectures and applications, such as OMG MASIF (mobile agent system interoperability facility)[4] and FIPA's agent-related specifications.[5] Further, the European network of excellence for agent-based computing (AgentLink),[6] international workshop series and conferences on the subject, such as ATAL,[7] CIA (Klusch and Kerschberg, 2000), Autonomous Agents, PAAM, and ICMAS, have all strongly pushed software agent technology since its public breakthrough around five years ago.

Intelligent agents for the Internet are commonly called information agents. But what exactly is an information agent? We define an *information agent* as an autonomous, computational software entity (an intelligent agent) that has access to one or more heterogeneous and geographically distributed information sources, and which pro-actively acquires, mediates, and maintains relevant information on behalf of users or other agents preferably just-in-time. Thus, an information agent is supposed to satisfy one or more of the following requirements:

- *Information acquisition and management.* It is capable of providing transparent access to one or many different information sources. Furthermore, it retrieves, extracts, analyses, and filters data, monitors sources, and updates relevant information on behalf of its users or other agents. In general, the acquisition of information encompasses a broad range of scenarios including advanced information retrieval in databases and also the purchase of relevant information from providers on electronic marketplaces.
- *Information synthesis and presentation.* The agent is able to fuse heterogeneous data and to provide unified, multi-dimensional views on information relevant to the user.
- *Intelligent user assistance.* The agent can dynamically adapt to changes in user preferences, information, and the network environment as well. It provides intelligent, interactive assistance for common users, supporting their information-based business on the Internet. In this context, the utilisation of intelligent user interfaces like believable, life-like characters can significantly increase not only the awareness of the user of their personal information agent but the way information is interactively dealt with.

Many (systems of) information agents have been developed or are currently under development in academic and commercial research labs, but they still have to wait to make

---

[4] http://www.fokus.gmd.de/research/cc/ecco/masif/index.html

[5] http://www.fipa.org

[6] http://www.agentlink.org

[7] http://www.atal.org

it out to the real world of Internet users broadly. However, the ambitious goal of satisfying all of the requirements mentioned above appears not to be very far away from accomplishment in the next ten years.

Information agents may be categorised into several different classes according to one or more of the following features (Klusch, 2000b):

1. *Non-cooperative* or *cooperative* information agents, depending on the ability of the agents to cooperate with each other for the execution of their tasks. Several protocols and methods are available for achieving cooperation among autonomous information agents in different scenarios, like hierarchical task delegation, contracting, and decentralised negotiation.

2. *Adaptive* information agents are able to adapt themselves to changes in networks and information environments. Examples of such kinds of agents are learning personal assistants on the Web.

3. *Rational* information agents behave in an economically rational way, based on utilitarian decision making. As self-interested agents they aim to increase their own benefits. One main application domain of such agents is automated trading and electronic commerce on the Internet. Examples include the variety of 'shopbots' and systems for agent-mediated auctions on the Web.

4. *Mobile* information agents are able to travel autonomously through the Internet. Such agents may enable, e.g., dynamic load balancing in large-scale networks, reduction of data transfer among information servers, applications, and migration of small business logic within medium-range corporate intranets on demand.

According to the definition and classification of information agents we can differentiate between communication, knowledge, collaboration, and rather low-level task skills as depicted in Figure 5.1. In this figure, the corresponding key enabling technologies are listed below each of the different types of skills. Communication skills of an information agent imply either accessing information systems and databases or processing input from human users or other agents. An agent naming service as well as an agent communication language (ACL) enable communication between intelligent agents on different levels. An ACL has to be considered rather on top of, for example, middleware platforms such as OMG's CORBA and Sun's Java RMI, or specific APIs such as JDBC (Java database connectivity), ODBC (open database connectivity), or OKBC (open knowledge base connectivity).

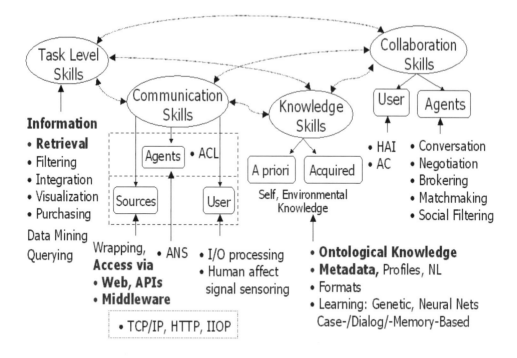

Figure 5.1 Basic skills of an information agent

Representing and processing of ontological knowledge and metadata, user profiles and natural language input, translation of data formats as well as machine learning techniques enable an information agent to acquire and maintain knowledge about itself and its user, the network, and the information environment. High-level collaboration of an information agent with other agents can rely on methods for, for example, service brokering, matchmaking, negotiation, and collaborative (social) filtering, whereas collaborating with its human users mainly implies the application of techniques stemming from human computer interaction and affective computing.

5.4   Agent-Based E-Trading

Despite the enormous potential of electronic commerce, sophisticated agent-based trading still remains a key challenge for economists, computer scientists and business managers. It might reshape the way we think about economic systems and business processes in an increasingly networked world.

In an open and increasingly commercialised cyberspace, personalised information agents not only may pro-actively discover and manage information relevant to its customers but also be paid and have to pay for any services they provide. Reasonably, trading information agents have to be equipped with effective and efficient methods for making economically rational decisions. This includes scenarios where rational information agents may (1) make purchases up to a pre-authorised limit, filter information and solicitation from vendors, (2) dynamically trade products and commodities such as even bandwidth and components within the business-to-business/consumer digital market, (3) decide on bids from service providers for its customer in some reverse auction online in a consumer-to-business e-commerce setting, and (4) increase the level of trust in their actions gradually over time, involving only manageable risks for both customers and vendors (Schillo, Funk and Rovatsos, 1999).

5.4.1   Basic enabling techniques for agent-mediated trading

In addition to the techniques for developing e-business solutions mentioned in section 5.2 many negotiation schemes and trading mechanisms for agent-based systems rely on:

- multi-attribute utility theory (Keeney and Raiffa, 1976), price comparison, content-based recommendation and user profiling, and blueprint learning of unknown Web pages,
- collaborative recommendation (Shardanand and Maes, 1995), coalition formation among autonomous agents (Kraus and Shehory, 1999), service matchmaking and brokering among trading agents (Sycara et al., 1998), voting (social choice), auction-based protocols, marketplaces (Wellman and Wurman, 1998), variations of the contract net protocol, and arbitration schemes in case of agents with conflicting interests (Tesch, Fankhauser and Ouksel, 2000).

While the first three of these mechanisms are mainly used in the domain of non-cooperative shopbots, the later are being used for multi-agent systems where different trading agents are provided with a common interaction (negotiation) protocol and individual strategies (Bouteillier, Shoham and Wellman, 1997). The choice of negotiation protocol depends on what properties the system designer wants the overall multi-agent system to have. Negotiation protocols for trading agents can be evaluated according to different criteria such as *social welfare* (the sum of all agents' payoffs in a given solution), *Pareto optimality* (there is no solution other than the negotiated one such that each agent is better off and no one is worse off), *individual rationality* (an agent's payoff in the negotiated solution is at least as high as it would get when not having participated in the negotiation), *stability* of the mechanism (no agent is better off behaving differently), as well as communication and computational efficiency.

During negotiation rational agents may perform reasoning using quantitative utility functions in ways well known from decision theory, especially in computing optimal choices and assessing their sensitivity to variations in probabilities. Qualitative information might provide constraints on preferences that guide this reasoning by selecting or constructing appropriate quantitative utility measures, though, such a construction is hardly explored to date. A variety of methods have been invented to allow agents to determine which negotiation strategy is more successful, for example, using Bayesian learning (Sycara and Zeng, 1997), evolutionary computing supported by case-based reasoning (Matos and Sierra, 1998), or fuzzy similarity rules (Sierra, Faratin and Jennings, 1999).

Many settings in which negotiations among trading agents take place imply decision-making under uncertainty. Approaches to deal with uncertain, partial, tentative or generic information within a negotiation usually make use of, for example, the notion of expected value of information (Howard, 1966) and utility (von Neumann and Morgenstern, 1944), as well as possibility theory (Dubois and Prade, 1986).

In the *expected utility theory* case, uncertainty is represented by probability distributions over a finite set C of consequences regarding a finite set S of situations and decisions (mapping S to C). In a given situation s, the agent looks up its real-valued utility function u (which models its preferences over C), and computes the global utility of a decision d (or action) as the expected value of its utility with respect to the corresponding probability:

$$EU_s(d) = \int_C u(c)\Pr(c \mid d)dc \approx \sum_{c \in C} u(c) \cdot \Pr(c \mid d) \qquad (5.1)$$

For this purpose the agent has to know the probability of each possible consequence c of a decision d in a given situation s. Rationally behaving agents are then assumed to always maximise expected utility. However, this approach fails to address making decisions in unforeseen circumstances or changing assumptions.

The *qualitative decision-theoretic* approach of using possibility theory replaces numeric preferences and probabilities with linear orderings presuming that the scales for preferences and belief have been interrelated and a decision-making policy, for example a minimax policy, has been adopted (Dubois et al., 1998). Uncertainty is represented by possibility distributions over C, an ordered uncertainty scale T, and a possibility distribution (mapping C to T) providing a plausibility ordering on C which can then be used to define qualitative utility functions. An example application of this type is given in Garcia et al. (1998). In this approach the refinement of bids is achieved for an agent (at an auction) by the use of individual information that is induced from a memory of cases composed from the history of past auctions.

After negotiation, the deals that the trading agents have agreed upon need to be executed in terms of exchanging traded goods and respective payments. This implies the problem of trust and traceability of the real-world parties associated with the trading agents. One option is to carry out such an exchange without enforcement by trusted third parties or relying on communities of trust. This is managed by intelligently splitting the exchange into smaller chunks for a safe sequence of deliveries and payments such that no agent is motivated to defect. At any point in the exchange, for both merchant and customer agent, the future gains from carrying out the rest of the exchange are larger than the gains from terminating the exchange prematurely. The latter can be tempting for a customer to do when, for example, the merchant at some point in the exchange has delivered more than the customer has yet paid for. However, if the chunks are interdependent in value, the sequencing cannot in general be done in polynomial time (Sandholm, 1997).

Another related issue concerns the policy of contract agreements among trading agents in cases when an agent wants to be freed from the obligations implied by a specific contract. This can be taken into account such that the mechanism of contracting itself allows for unilateral decommitments at any point in time by specifying penalties to be paid by an agent to the other party of the contract. Such levelled commitment contracts have proven to outperform the full commitment protocol by increasing both contract parties' expected payoffs and enabling contracts in settings where no full commitment contract is individually rational for each of the parties (Sandholm and Lesser, 1996). A closely related approach for relaxing the full binding of classical contracts is to formalise and reason on conditional obligations. This corresponds to performing argument-based reasoning to identify explicit reasons for alternatives and generic preferences of contracts which may be compared with importance measures (van der Torre, 1997).

In the following sections we briefly sketch the techniques of content-based and collaborative recommendation, coalition formation, and auctions. A more in-depth survey of techniques for automated, distributed rational decision-making of self-interested agents can be found in, for example, Sandholm (1999).

5.4.1.1   Content-based and collaborative recommendation

Collaborative recommendation (or social filtering) focuses on identification of similar users and their opinions to recommend items (Good et al., 1999; Schafer, Konstan, and Riedl, 1999). It is a powerful method for leveraging the information contained in user profiles. In contrast to content-based filtering through which items are recommended to a user according to correlations found between the items' descriptions (treated as text documents via traditional information retrieval techniques) and the given or observed users preferences stored in a profile, an agent rates items chosen by its user and compares the corresponding user preference vector to that of other known users. It then recommends other items that have been recommended by users who share similar likes *and* dislikes. For this purpose it has to collaborate with other agents to gain the respective knowledge.

Collaborative recommendation essentially automates the process of "word of mouth" in a given community and currently is one of the prevalent techniques for e-commerce targeting a special user community via a password-protected multiagent system such as Firefly or GroupLens (Resnick et al., 1994). In addition, trust among users and agents is even easier to gain, since it is very difficult to manipulate the recommendations an agent makes to its user via social filtering.

A common techniques to find similar users and predict weighted averages of user ratings is to determine the correlation between the user's preference vectors by using minimum square error, or the Pearson correlation:

$$\sum_{i\in\{1,\ldots x-1,x+1,\ldots n\}}[\vec{r}_{+}(u_{x})-\vec{r}_{-}(u_{x})]\cdot[\vec{r}_{+}(u_{i})-\vec{r}_{-}(u_{i})] \tag{5.2}$$

$\vec{r}_{+/-}(u_{k})$ positive/negative ratings of user k

However, traditional collaborative recommendation has a few shortcomings. Initial users can bias ratings of future users, different (context-based) points of view of users are not taken into account, and there is no learning from negative cases involved. These drawbacks are significantly different from those of content-based filtering still used by many shopbots to date. In particular, content-based methods cannot help to recommend more of what the user has seen before and liked, and no assessments of style, quality, etc, can be offered to the

the user. Ongoing work on integrating content-based and collaborative filtering are reported, for example, in the RAAP project (Delgado, Ishii, and Ura 1998).

5.4.1.2   Coalition formation

Self-interested, autonomous agents may negotiate rationally to gain and share benefits in stable (temporary) coalitions. This is to save costs by coordinating activities with other agents. For this purpose, each agent determines the utility of its actions and productions in a given environment by an individual utility function. The value of a coalition among agents is computed by a commonly known characteristic function that determines the guaranteed utility the coalition is able to obtain in any case. In a characteristic function game, the agents may use imposed individual strategies to achieve a desired type of economically rational behaviour, such as altruistic, bounded rational, or group rational. In any case, the distribution of the coalition's profit to its members is decoupled from its obtainment, but is supposed to ensure individual rational payoffs to provide a minimum of incentive for the agents to collaborate.

The formation of stable coalitions relies on derived concepts from cooperative game theory, economics, and operations research. It covers two main activities: (1) the formation of coalition structures, that is, partitioning the set of agents into coalitions, so as to maximise the monetary value for accomplishing tasks in individual or group rational coalitions, and (2) the distribution of gained benefit among coalition participants. These activities may be interleaved and are not independent.

Interesting cases are those where coalition formation is concerned with non-superadditive environments where at least one pair of potential coalitions is no better off by merging into a coalition caused by factors such as communication and coordination overhead costs, decrease of coalition value as a result of restricting utility constraints posed by agents joining a coalition, or anti-trust penalties for specific coalitions (Kraus and Shehory, 1999).

The meaning of stability of a formed coalitions relies on the chosen game-theoretic concept for payoff division within coalitions. For example, the payoff division may be defined by the Shapley value, the Core, the Bargaining Set, or the Kernel (Kahan and Rapoport, 1984). Searching for an optimal coalition structure (given a set A of agents) among the exponential number of $|A|^{|A|/2}$ possible coalition structures is computationally hard since one has to try at least $2^{|A|-1}$ coalition structures (Sandholm et al., 1998, 1999).

In environments where published interests and utilities used for coalition-forming negotiation cannot be verified (most current protocols do not tackle fraud caused by different types of lies), arbitration schemes for competing agents with conflicting interests may help to circumvent such situations (Tesch and Fankhauser, 1999).

Although well-grounded techniques for coalition formation among self-interested agents are known, none of them have been used so far in the public Web. Current application domains include, for example, cooperative information systems and decentralised power transmission planning (Contreras et al., 1999). A publicly available simulation environment for coalition formation based on selected coalition theories is provided by COALA.[8]

---

[8] http://www.dfki.de/~klusch/COALA/

Further research issues include, for example, efficient methods for dynamic formation of multiple overlapping coalitions, levelled trust, and stability. *Dynamic coalition formation* (DCF) theory deals with the problem of scenarios in which (1) agents may leave or enter the negotiation process at any time, and (2) utilities computed over a continuous stream of tasks to accomplish may rapidly change and are not necessarily completely known to the agents (Klusch, 2000b). Classical game-theoretic notions of coalition stability and respective negotiation algorithms are not applicable in such settings. Scenarios inducing uncertain, time-limited, context-based utilities and coalition values exacerbate the problem of DCF. For example, an agent may determine the degree of membership to potential coalitions based on bargaining and the possible level of its commitment indicating the degree of collaboration that it desires. Related work on fuzzy coalition forming (Mares, 1997; Ketchpel, 1994; Aubin, 1981), rational revision of preferences, and other qualitative approaches to decision making based on partial, uncertain, and tentative information hold promise to be useful for coping with these issues of the DCF problem.

DCF methods could be applied to, for example, formation of temporary customer coalitions on the fly for bidding, purchasing, and sharing an appropriately partitioned set of items available at multiple auction houses. In the work of Sandholm and Huai (2000) this is intended to be supported by mobile agents in the future.

5.4.1.3   Auctions

Auctions theory (Wolfstetter, 1996) analyses protocols and agents' strategies in auctions. An auction is a price-fixing mechanism in which negotiation is subject to a very strict coordination process. It consists of an auctioneer who wants to mediate the exchange of given items between buyers and vendors for sale at the highest possible price, and potential bidders who want to buy them at the lowest possible price. Asynchronous bidding mechanisms are mostly based on open-outcry with price changes, or sealed bids with periodic partial revelation mechanisms.

Online auctions appear to be unnecessarily hostile to customers due to the 'winner's curse' and offer no long-term benefits to merchants. If bidders have reasonable information about the worth of the item, then the average of all the guesses is likely to be correct. However, the winner who offered the bid furthest from the actual value, thus, pays more for the item than its value, so any auction is basically a win lose game.

Any auction may be classified along three dimensions of (1) bidding rules, including, for example, bid format; and many-to-one or many-to-many participation; (2) clearing policy, such as pricing, clear schedule and closing; and (3) information revelation policy, including price quotes, quote schedule, etc. Prominent protocols for single-item auctions include:

- first-price, open-cry, so-called *English auction*. The dominant strategy for consumers here is to bid up to their true, maximum value.
- Descending-price, open-cry, so-called *Dutch auction* that guarantees the auctioneer the purchase of items at the highest possible price.
- *first-price, sealed-bid auction* where each bidder submits one bid in ignorance of all other bids. The highest bidder wins and pays the amount he or she bid. This has the potential to force buyers and seller into price wars since the sealed bid of any bidder depends on what s/he believes of all other opponents' bids.

- second-price, sealed-bid, so-called *Vickrey auction* where the winning bidder pays the price of only the second highest bid (Sandholm, 1996).

Auctions for multiple (identical or heterogeneous) items for sale include the discriminatory, the double, and the matrix-type auctions. Further auction-based mechanisms are discussed in Fischer, Russ and Vierke (1998).

In a *discriminatory auction* sealed bids on multiple items are sorted from high to low, and items are awarded at highest bid price until the supply is exhausted. Winners pay exactly what they bid but, of course, usually different prices for different items.

As in a two-sided auction where bids and asks are allowed, the *double auction* (Wurman, Walsh and Wellman, 1998) enables both sellers and buyers to submit bids for single or multiple units of items which are then ranked highest to lowest to generate demand and supply profiles. From the profiles, the maximum quantity exchanged can be determined by matching selling offers (starting with lowest price and moving up) with demand bids (starting with highest price and moving down). This format allows buyers to make offers and sellers to accept those offers at any particular moment. A *continuous double auction* of multiple items is an auction where many individual transactions are carried out at any single moment and trading does not stop as each auction is concluded (Preist and van Tol, 1998).

Under the assumption of subjective private value, all four basic auction types listed above can be shown to yield the same expected price and revenue to the seller when bidders are not risk-averse but risk-neutral and symmetric (meaning they use the same measurements to estimate their valuations). This implies that auction choice is not crucial because each format yields on average the same payoff. But revenue equivalence does not hold true under the common-value assumption (when bidders have similar evaluations).

Bidding is often the result of correct predictions about the behaviour of others and sometimes that means guessing the extent of someone else's information correctly. Information agents may provide background knowledge and additionally acquired information on competitors concurrently to support the bidding of its user, just in time.

Another issue concerns shills, which means agents are planted by the auctioneer to manipulate the valuation of the auctioned good by raising bids to stimulate the market. The same goes the other way round with formation of coalitions of buyers who agree not to outbid one another but distribute the purchased items among themselves privately. Both issues are considered illegal but are hard to detect; thus, mechanisms to reduce the agents' incentive for both types of action *a priori* have to be embedded into the negotiation protocol explicitly or indirectly as part of the protocol's theoretical features.

5.4.2   The non-cooperative case: shopbots

Quite popular but essentially very simple examples of non-cooperative rational information agents are shopbots on the Web. Examples include mySimon.com, Junglee/Yahoo!, Jango/Excite, shopfido.com, compare.net, and evenbetter.com. Shopbots do not sell any product, but guide the customer to recommended merchant's online stores offering the required items. The recommendation is determined either by comparing prices of products or by applying multi-attribute utility theory (Keeney and Raiffa, 1976; Fishburn, 1977).

The latter theory provides a way of representing and calculating the utilities of action outcomes by decomposing the utilities into formulations of the value-relevant factors that make up the outcomes of recommended actions. Recommendation is determined with fixed and defeasible (non-monotonic) assumptions about the independence of the values assigned to different factors and the magnitude of these values. In practice, this theory offers methods for composing utility measures and construction of libraries of standard forms for utility functions.

The first generation cross-merchants comparison shopping agents, such as BargainFinder of Andersen Consulting, are limited to comparing merchant offerings based only on price. In contrast, second-generation shopbots, such as frictionless.com recommend product items based on evaluation of multiple attributes such as price, quality of desired product, delivery times and costs, return policies, promotions and gift services as well as customer support and reputation of merchants. The underlying integrative negotiation analyses the decision problem of what and who to buy from (in terms of the transaction) qualitatively using attribute constraints, and solves this problem by applying finite-domain constraint satisfaction techniques (Tsang, 1993).

The underlying assumptions for both generations of shopbots are that (1) merchants reveal the relevant information of their product items and business policies to the agent, and (2) the content of the respective Web pages can be automatically scanned and understood by the visiting shopbot. However, dynamically created Web pages and information encoded in graphics, video, and audio format pose challenges to any of currently deployed shopbots.

A theoretical study has been conducted by Greenwald et al. (2000) on the potential impact of widespread shopbot usage and the price dynamics that may ensue from various mixtures of automated pricing agents, so-called pricebots, that employ price-settings algorithms in an attempt to maximise their vendor's profits. Among other issues noted, it has been suggested that price wars may appear and game-theoretic equilibria can dynamically arise. However, it remains to be seen if shopbots can successfully compete with large-portal sites such as amazon.com or barnesandnoble.com, and online retail auctions such as eBay.com in the long run.

The rise of synthetic characters and avatars on the Web (Elliott and Brzezinski, 1998) such as Cyberella of DFKI GmbH (Andre, Rist and Mueller, 1998) or Cor@ at the Web portal of the Deutsche Bank, not only holds great promise for increasing users' awareness and acceptance of personalised intelligent assistants for improving convenient guidance through online stores, but significantly increases the companies' expected profits. A lot of life-like synthetic characters and personal assistants have been developed by, for example, Extempo Inc., DFKI, Microsoft, and MIT Media Lab. Such characters are programmable to behave in accordance to different personality traits as specified by the developer in the context of the considered e-business application. For example, a set of different characters at a car dealer's Web portal can simulate personalised sales dialogues between a virtual vendor and potential customers. Each of these characters adopts a different role, trying to persuade the real customer to purchase an advertised item through role-based argumentation in a simulated conversation. However, one has to be aware that a rational information agent can present itself to the user in the form of such a believable synthetic character as a kind of alter ego but should never be identified with it, as it is just its intelligent user interface. One further step would be to inhabit 3D shopping malls on the

Web such as culthouse.de or vira.de with 3D information agents assisting users on their virtual shopping tours. First attempts in this direction have been reported recently in the context of digital cities (Ishida and Isbister, 2000).

Another species of single agents in the e-business domain is that of chatter-bots which use low-complexity case-based reasoning techniques (Lenz et al., 1998) to guide users through Web-based product catalogues or to answer the 20% of questions that generate 80% of call volume to customer-service call centres.

5.4.3   The collaborative case: agents on markets and auctions

Agents may collaborate for gaining or sharing benefits, though the degree of rationality or collaboration depends on the outcome of the negotiation. Such distributed negotiation among agents may impact, for example, the charges for services provided as well as the kinds of services or goods themselves. Negotiation can take place with or without any mediator agent and usually follows some given conventions on communication such as, what ontology and language to use for this purpose, the social behaviour of the agents in a system, and each agent's individual rights, obligations, and commitments. Free markets and auctions are the commonest virtual institutions for e-commerce to be mediated by collaborating rational information agents; they are the means for C2C and B2C e-commerce, respectively.

Marketplaces (Wellman and Wurman, 1998) provide locations where multiple information agents of customers and vendors may meet to negotiate and exchange relevant data and information. Negotiation concerns include the amount to charge or to pay for services as well as the kinds of services or goods themselves.

Almost all current Internet auction types are single-resource, one-sided, and not executed in real time in a strong sense. Future trends include combinatorial auctions with lower and upper prices for product bundles as well as reverse auctions where service providers bid to satisfy some customer's request for a kind of service.

Prominent   examples   for   agent-based   marketplaces   and   auctions   are Kasbah/MarketMaker (Chavez et al., 1997)[9], ZEUS virtual marketplace (Collis and Lee, 1998), AuctionBot,[10] UMDL (Durfee, 1999), and FishMarket.[11]

UMDL (University of Michigan Digital Library) is an agent-based digital library offering electronically available information content and services in a distributed environment. It relies on a multi-agent infrastructure (the service market society, SMS) with agents that buy and sell services from each other using a given set of commerce and communication protocols. Within the SMS, self-interested agents are able to find, work with, and even try to outsmart each other, as each agent attempts to accomplish the tasks for which it was created. Learning in the context of SMS provides a way for agents in the SMS to develop expectations and strategically reason about others, and exploit these expectations to their mutual benefit.

---

[9] http://ecommerce.media.mit.edu/{Kasbah/kasbah.html, maker/maker.htm}

[10] http://auction.eecs.umich.edu

[11] http://www.iiia.csic.es/Projects/fishmarket/newindex.html

Kasbah/MarketMaker is a simple agent-based marketplace which has been developed at MIT Media Lab. Trading of goods is performed among buyer and seller agents on a central marketplace using a simple language for advertisements and requests. Each agent has knowledge about the (type of) good that it has to buy or sell by pro-actively seeking out potential best deals and negotiate them on their user's behalf. These deals are subject to user-specified constraints in terms of desired price, lowest (or highest) acceptable price, a date to complete the deal, and one of three pre-defined types of price decay functions. These linear, quadratic, and exponential functions correspond to greedy, moderate, and anxious behaviour of buyers or sellers, respectively. Upon completion of a deal (and consequent transaction) both parties are able to rate the other party's part of the deal in terms of, for example, product quality and timely completion of transaction. Agents may use these ratings to determine their willingness to follow up a negotiation with agents whose users do not match a given reputation threshold.

5.5   Conclusions and Outlook

Using autonomous trading agents may have different impacts on Internet-based economies and business in the future. Such agents may make purchases up to an authorised limit, filter information and solicitation from different merchants, and dynamically trade any type of good pro-actively on behalf of its users in markets and auctions. For this purpose a variety of basic enabling techniques are available. Ubiquitous electronic marketplaces rationally brokered by heterogeneous, intelligent and life-like agents are also driving research and development of related technologies such as generation and (re-)use of ontologies for electronic commerce and the integration of mobile telecommunication and the Internet.

Mobile commerce supported by personalised, rational information agents residing on WAP-enabled access devices such as pagers, organisers, (sub)notebooks, or UMTS cell phones, is still a vision for the common Internet user but is not too far away from realisation.

Other future domains of agent-mediated electronic business are integrated commerce that encompasses agent-based coordination of the whole supply chain associated with any product ordering by a customer, and collaborative customer relationship management. Smaller and buyer-centred niche markets will become viable with more diverse goods, content, and services, with personalised agents guiding loyal customers.

Currently deployed agents are able to, for example, notify users on stock prices, make intelligent recommendations, and negotiate in different settings on virtual auctions and marketplaces, though most of the approaches reported in this survey are not practical systems yet. Agent technology has just begun to be considered influential in terms of thinking of a new decentralised digital economy, but still fails to provide easy-to-use, secure concepts and standards for the public mass market including both online vendors and customers. It remains to be seen to what extent different types of trading agents in which domain of e-business will play a major role.

5.6   References

Andre, E., Rist, T. and Mueller, J. (1998) WebPersona: A life-like presentation agent for the world-wide Web, *Knowledge-based Systems*, vol. **11(1)**.

Aubin, J. P. (1981) Cooperative fuzzy games, *Mathematics of Operations Research*, vol. **6(1)**.

Bouteillier, C., Shoham, Y., and Wellman, M. (1997) Economic principles of multi-agent systems, *Artificial Intelligence*, vol. **94**.

Chavez, A., Dreilinger, D., Guttman, R. and Maes, P. (1997) A Real-Life Experiment in Creating an Agent Marketplace. *Proc. Intl. conference on Practical Application of Multiagent Systems PAAM, London, UK.*

Collis, J. C., and Lee, L. C. (1998) Building electronic marketplaces with the ZEUS agent tool-kit, In Proc. AMET Workshop, P. Noriega and C. Sierra (eds.), *Agent-Mediated Trading*, LNAI series, vol. **1571**, Springer.

Contreras, J., Klusch, M., Vielhak, T., Yen, J. and Wu, F. (1999) Multi-agent coalition formation in transmission planning – bilateral Shapley value and Kernel approaches, *Proc. 13th Intl. Power Systems Computation Conference PSCC, Trondheim, Norway.*

Delgado, J., Ishii, N. and Ura, T. (1998) Content-based collaborative information filtering – Actively learning to classify and recommend documents, In *Proc. CIA-98 Workshop*, M. Klusch, G. Weiss (eds), *Cooperative Information Agents III*, LNAI series, vol. **1435**, Springer.

Dubois, D. and Prade, H. (1986) *Possibility theory: An Approach to Computerised Processing of Uncertainty*, New York, USA, Plenum.

Dubois, D., Le Berre, D., Prade, H. and Sabbadin, R. (1998) Logical representation and computation of optimal decisions in a qualitative setting, In *Proc. of 15th National conference on Artificial Intelligence AAAI*, Menlo Park, CA, USA, AAAI Press.

Durfee, E. (1999) Strategic reasoning and adaptation in an information economy, In M. Klusch (ed.), *Intelligent Information Agents*, Springer.

Elliott, C. and Brzezinski, J. (1998) Autonomous agents as synthetic characters, *Artificial Intelligence Magazine*, vol. **19(2)**.

Fischer, K., Russ, C. and Vierke, G. (1998) Decision theory and coordination in multiagent systems, DFKI Research Report RR-98-02, German Research Centre for AI.

Fishburn, P. C. (1977) Multiattribute utilities in expected utility theory, In *Conflicting Objectives in Decisions*, D.E. Bell et al. (eds.), NY, John Wiley & Sons.

Garcia, P., Gimenez, E., Godo, L. and Rodriguez-Aguilar, J. A. (1998) Bidding strategies for trading agents in auction-based tournaments, In *Proc. AMET Workshop*, P. Noriega and C. Sierra (eds.), *Agent-Mediated Trading*, LNAI series, vol. **1571**, Springer.

Good, N. et al., (1999) Combining collaborative filtering with personal agents for better recommendations. *Proc. Conference of the American Association of Artifical Intelligence (AAAI-99), July.*

Greenwald, A., Kephart, J. O. and Hanson, J. (2000) Dynamic pricing by software agents, *Computer Networks*, Elsevier.

Gruber, T. (1992) Ontolingua: A mechanism to support portable ontologies, Stanford University, Technical Report KSL-91-66, 1992. See also: A translation approach to portable ontology specifications, *Knowledge Acquisition*, vol. **5(2)**, 1993

Guttman, R., Moukas, A. and Maes, P. (1999) Agents as mediators in electronic commerce, In M. Klusch (ed.), *Intelligent Information Agents*, Springer.

Howard, R. A. (1966) Information value theory, *IEEE Trans. on Systems Sciences and Cybernetics*, vol **2**.

Ishida, T. and Isbister, K. (eds.) (2000) Digital Cities – *Technologies, Experiences, and Future Perspectives*, LNCS series, vol. **1765**, Springer.

Jennings N. R. and Wooldridge, M. (eds.) (1998) *Agent technology*, Springer/Unicom.

Kahan, J. P. and Rapoport, A. (1984) *Theories of Coalition Formation*, Hillsdale NY, Lawrence Erlbaum Associates.

Keeney, R. and Raiffa, H. (1976) *Decisions with Multiple Objectives: Preferences and Value Tradeoffs*, John Wiley & Sons.

Ketchpel, S. (1994) Forming coalitions in face of uncertain rewards, *Proc. National conference on Artificial Intelligence (AAAI)*.

Klusch, M. (2000a) Information Agent Technology for the Internet: A Survey, In D. Fensel et al. (eds.), *Data and Knowledge Engineering*, Special Issue on Intelligent Information Integration.

Klusch, M. (2000b) The future of information agents in cyberspace, M. Klusch (ed.), *Intelligent Cooperative Information Systems, Special Issue on Intelligent Information Agents: Theory and Applications 1*, World Scientific, to appear.

Klusch, M. and Kerschberg, L (eds.) (2000) *Cooperative Information Agents IV*, LNAI series, vol. **1860**, Springer.

Kraus, S. and Shehory, O. (1999) Feasible formation of coalitions among autonomous agents in non-superadditive environments, *Computational Intelligence*, vol **15(2)**.

Lenz, M. et al., (1998) *Case-based Reasoning Technology*, LNAI series, vol. **1400**, Springer.

Manchala, D. W. (2000) E-commerce trust metrics and models, *IEEE Internet Computing*, vol **4(2)**.

Mares, M. (1997) Fuzzy coalition forming, *Proc. of Intl. conference IFSA*.

Matos, N. and Sierra, C. (1998) Evolutionary computing and negotiating agents, In *Proc. AMET Workshop*, P. Noriega and C. Sierra (eds.), *Agent-Mediated Trading*, LNAI series, vol. **1571**, Springer.

Merz, M. (1999) Electronic commerce, dpunkt.

Noriega, P. and Sierra, C. (eds.) (1999) Agent mediated electronic commerce, *Proc. AMET-98 Workshop on Agent-Mediated Electronic Commerce*, LNAI vol. **1571**, Springer.

Preist, C. and van Tol, M. (1998) Adaptive agents in a persistent shout double auction, In Proc. 1st *Intl. conference on Internet, Computing, and Economics*, ACM Press.

Resnick, P. et al. (1994) GroupLens: An open architecture for collaborative filtering of netnews, In *Proc. of Computer Supported Cooperative Work (CSCW)*, NY, USA.

Sandholm, T. (1996) Limitations of the Vickrey Auction in Computational Multiagent Systems, *Proc. of Intl. conference on multiagent systems (ICMAS)*.

Sandholm, T. (1997) Unenforced e-commerce transactions, *IEEE Internet Computing*, vol. **1(6)**.

Sandholm, T. (1999) Distributed rational decision making. In G. Weiss (ed.), *Multiagent Systems*, MIT Press.

Sandholm, T and Huai, Q. (2000) Nomad: Mobile agent system for an Internet-based auction house, *IEEE Internet Computing*, vol. **4(2)**.

Sandholm, T., Larson, K. S., Andersson, M., Shehory, O. and Tome, F. (1998) Anytime coalition structure generation with worst case guarantees, In *Proc. of National conference on Artificial Intelligence AAAI, Madison, USA*.

Sandholm, T. and Lesser, V. (1996) Advantages of levelled commitment contracting protocol, In *Proc. National conference on Artificial Intelligence AAAI, Portland, USA*.

Sandholm, T. et al. (1999) Coalition structure generation with worst-case guarantees, *Artificial Intelligence*, vol. **111 (1-2)**.

Schafer, J. B., Konstan, J., Riedl, J. (1999) Recommender systems in e-commerce, *Proc. ACM Conference on Electronic Commerce*.

Schillo, M., Funk, P. and Rovatsos, M. (1999) Who can you trust: Dealing with deception, In R. Falcone (ed.), *Proc. Workshop on Deception, fraud, and trust, Intl. conference on Autonomous Agents*.

Shardanand, U. and Maes, P. (1995) Social information filtering: Algorithms for automating 'word of mouth', In *Proc. Computer-Human Interaction conference (CHI), Denver, USA*.

Sheth, A., Illaramendi, A., Kashyap, V. and Mena, E. (1996) Managing multiple information sources through ontologies: Relationship between vocabulary heterogeneity and loss of information, In *Proc. European conference on AI (ECAI)*.

Sierra, C., Faratin, P. and Jennings, N. (1999) Deliberative automated reasoning using fuzzy similarities, *Proc. EUSFLAT-ESTYLF, Joint conference on Fuzzy Logic*.

Steels, L. (1998) The origins of ontologies and communication conventions in multiagent systems, *Autonomous Agents and Multi-Agent Systems*, vol. **1(2)**.

Sycara, K. and Zeng, D. (1997) Benefits of learning in negotiation, *Proc. National conference on Artificial Intelligence AAAI, Providence, USA*.

Sycara, K., Lu, J., Klusch, M. and Widoff, S. (1999) Dynamic service matchmaking among agents in open information environments, ACM SIGMOD Record, Special Issue on Semantic Interoperability in Global Information Systems. Long version available as technical report CMU-RI-TR-98-22, Carnegie Mellon University, Pittsburgh, USA.

Tambe, M., Pynadath, D. V., Chauvat, N. (2000) Building dynamic agent organizations in cyberspace, *IEEE Internet Computing*, vol. **4(2)**.

Tesch, T and Fankhauser, P. (1999) Arbitration and matchmaking for agents with conflicting interests, In M. Klusch, O. Shehory, G. Weiss (eds.), *Cooperative Information Agents III, Proc. CIA 1999*, LNAI series, vol. **1652**, Springer.

Tesch, T., Fankhauser, P. Ouksel, A. (2000) Arbitration protocols for competing software agents, *Intelligent Cooperative Information Systems*, M. Klusch (ed.), Special Issue on Intelligent Information Agents: Theory and Applications 1, World Scientific, to appear.

Tsang, E. (1993) *Foundations of Constraint Satisfaction*, Academic Press.

van der Torre, L. W. N. (1997) Reasoning about obligations: Defeasibility in preference-based deontic logic, *Tinbergen Institute of Research Series*, vol. **140**, Amsterdam, Thesis.

von Neumann, J. and Morgenstern, O. (1944) *Theory of Games and Economic Behaviour*, Princeton University Press, Princeton, NJ, USA.

Wellman, M. and Wurman, P. R. (1998) Market-aware agents for a multiagent world, *Robotics and Autonomous Systems*, vol **24**.

Wolfstetter, E. (1996) Auctions: An introduction, *Economic Surveys*, vol. **10(4)**.

Wooldridge, M. (1999) Intelligent agents, In G. Weiss (ed.), *Multiagent Systems*, The MIT Press.

Wurman, P. R., Walsh, W. E. and Wellman, M. P. (1998) Flexible double auctions for electronic commerce: Theory and implementation, *Decision Support Systems*, **24**, 17-27.

6

Distributed Control of Connection Admission to a Telecommunications Network: Security Issues

J. Bigham, A. L. G. Hayzelden, J. Borrell, S. Robles

6.1   Introduction

Bigham et al. (1999 a,b) describe a distributed architecture that provides decentralised run-time management of connection to a telecommunications network. The architecture supports competing Service Providers (SP) and a competing Network Provider (NP). The architecture has been designed as a multi-agent system. This agent system manages connections in a real network of commercially available switches transferring different applications such as video. The multi-agent system was shown to provide more flexibility than conventional signalling approaches. It also provides an open framework for the selection of the service provider to carry the connection at the required quality of service and for the deployment of network control schemes. One of the system's major contributions is the ability to allow different service providers to use different control algorithms on the same logically partitioned physical infrastructure. It is believed that such a structure is appropriate for networks where competing SPs are offering service.

Nevertheless, all security aspects relating to the auction process were ignored. It was assumed that an underlying security infrastructure would handle such problems. Since some security aspects are also present in high layers of the architecture, this assumption is not enough to solve all the security problems. The aim of this chapter is to propose an underlying security infrastructure and then use it to solve the security problems of the system described in Bigham (1999b). This is hopefully of interest to anyone constructing an auction system for a similar application. Our experience working in multi-agent systems' security applied to the design of an Electronic Voting Scheme (Riera and Borrell 1999) is used here.

In section 6.2 the auction process and its context are described. In section 6.3 security problems of the scheme are described and illustrated. In section 6.4 we present the security mechanisms that will be used in section 6.5 to solve problems described in section 6.3. Finally, conclusions of this chapter are described in section 6.6.

6.2   The Auction Mechanism for Selecting an SP

We have chosen what is expected to be a common business model for selling bandwidth to illustrate the approach. In the management plane a network provider manages an entire physical network. The NP leases the use of bandwidth to SPs which are in competition with each other and, except for the SP owned by the NP, the NP.

The Connection Agent (CA) conducts a first-price sealed-bid auction with the appropriate Resource Agents (RAs) for user-requested connections. It gets replies from relevant RAs and it then decides on the preferred service provider or the preferred offer from that service provider in the case of multiple offers and instructs the chosen RA to install the connection. The chosen RA then interacts with the Switch Wrappers.

Immediate problems of scalability arise through having one CA, and these are addressed later. However, we first consider the reasons for having a CA at all. If we considered an architecture without a CA or, what is the same, a CA owned by the user, we would have to concern ourselves with direct interactions between users and RAs. Although this is feasible, it is not desirable to have such a structure. The Proxy user agent interacts between the CA and the end user.

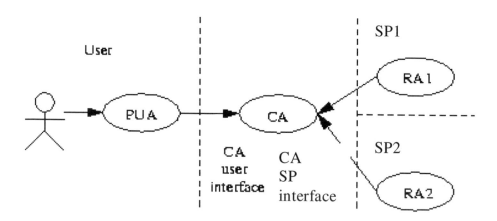

Figure 6.1 Ownership of system elements

Figure 6.1 shows ownership areas for some system elements. For some elements ownership is clear: the user agent PUA_n is owned by user_n, and RA_n is owned by NP_n, as the agents act on their behalf by definition. Ownership of the CA is not as clear: it could be owned by the user (in fact, we could then have a system without a CA, since its functionality could be integrated into the PUA) or by a third party.

One of the benefits of the CA is to offer to the users a true image of the SPs. It is entirely objective. An NP or SP may use publicity, rumours, rival campaigns to gain prestige and to benefit from past history. In this way the user's image of an NP or SP may be false. The CA gives the users a real present image of the SP through neutral auctions. However, it is still up to the PUA to handle assessments of reliability and to balance such attributes and perhaps other less tangible attributes against price.

Once we have a PUA→CA interface and an RA→CA interface, the system representing the CA can be implemented in several ways. This helps to add scalability to the scheme in a manner that is transparent to a user and an SP. The CA will offer both user and SP interface standards. If an upgrade is needed, such as adding a new SP interface standard, the changes to the CA should not be noticeable to the user. If the CA were not present, then each PUA would need to alter its SP interface.

Both, users and the SPs, need a trusted party to perform trading. Otherwise users may be sceptical in using an electronic scheme to choose their SP (in the introduction of Riera (1999) the problem of user scepticism and trust is discussed). In order to improve trust worthiness we present the Regulatory Board in section 6.5. We can consider the CA as a real auction house, trying to contact buyers and sellers while vouching for fair trading conditions and objectivity.

Other structures than a simple auction, such as a quote-driven market system as used in commodity markets are possible but are not analysed here.

6.3   Security Problems

We now analyse security problems within the scheme described. Several are well known, but enumerated for completeness. After each problem is described, ways of exploiting the lack of security are shown. In most cases it is shown how unfair benefits can be extracted from the system. We have sorted these problems into three categories: problems of authentication and secrecy, problems related to the bidding protocol and problems of scalability and reliability.

6.3.1   *Authentication, secrecy and trust*

6.3.1.1   Authentication PUA↔CA
A user (in this case through his or her PUA) must know if the connection is established with the real CA that has been assigned to him or her. On the other hand, only authenticated users may be allowed to start the connection mechanisms to the CA.

Without this two-way authentication, many attacks are possible. First, if no PUA←CA authentication exists (i.e. the PUA cannot be sure of the CA's identity), attacks against users can succeed. An attack can occur if an illegal party pushes in between the PUA and CA, pretending to be the CA for the PUA. Then, it can manipulate information to benefit the party inserting the pretender. Secondly, if no PUA→CA authentication exists attacks from anonymous users trying to collapse both the CA and NP or SP are not punishable.

6.3.1.2   Authentication RA↔CA

The RA sends information to and receives information from the CA. The RA has to be sure of the CA's identity, since a false CA could arrange connections for non-existent users asking for particular capacities and QoS, hence essentially interrogating the RA. On the other hand, when an RA enters into the system, the CA has to be sure that it is a legal RA acting on behalf of a legal SP or NP. The CA is acting, in part, on behalf of the user, and so should not offer invalid connections.

An attack consists of an NP inserting an illegal RA into the system, offering false connections. If the NP cannot offer a bid due to too high utilisation, its allied RA can make a false bid and then renege. So the auction will start again and perhaps this time the NP will have capacity. In the other hand, this NP could also insert a false CA. This CA that is allied to the NP could again interrogate and send spurious requests and so confuse the RAs by making them offer connections for non-existent users.

6.3.1.3   Insecure communication channels

None of the communications between agents in the system are secret. In particular, RA↔CA communications are observable by any third party. This is a problem, as anybody (e.g. the NP hosting the CA) can listen to the communication channels and act depending on the traffic seen. Even manipulation of data is possible.

Attack is easy by any RA. If a particular RA, say RAj, listens to all bids offered by all other RAs then it can send a last bid marginally better than the rest and so be rewarded. For instance, offering *{ Min(bid(RAi)) [for all i] } less epsilon* will automatically result in being awarded the bid.

6.3.1.4   Certification of the CA

Before a user connection request is sent, the user has to be sure that the auction in the CA will follow some known rules and that it will not favour any SP. An untrustworthy CA could let a malicious NP replace code in the CA to help its own SP to be awarded with a connection in the auction phase. Also, when a new RA enters into the system, acting on behalf of an SP, the SP needs to be sure that it will have the same opportunities to be awarded with connections in the auction phase as other RAs, i.e. the code executing the CA is the right one. An SP's RAs may refuse to enter the system when in doubt.

We need a trust model for the CA to avoid attacks against code in the CA. A way to achieve this could be to certify the code of the CA.

A problem that now arises is deciding who should certify the CA code. All parts in the systems must trust in the CA, so this is a sensitive problem. If the certifying authority may be corrupted the problem is bigger, simply because all parties will believe the CA is valid when it is not.

6.3.2   Bidding protocol

6.3.2.1   Same opportunities for all bidders

Each RA in the system has to have the same opportunity to be awarded a connection in the auction phase, independent of the distance from the CA or the connection type.

A problem can arise when an RA is located far from the CA. In this case it will be harder for the RA to be awarded because of awarder timeouts. If a different RA bids the same offer, and because of its shorter distance from the CA that bid arrives within the time frame, it will be awarded.

6.3.2.2   Denial of service
Since all bids are sent to the CA in a one-way protocol we can imagine a denial of service attack. An RA can prevent some or all bids sent by other RAs from arriving at the CA. This can be done by corrupting the network data or keeping the network busy (this is possible in many networks). Doing this, this RA can easily be awarded the connection by the CA.

6.3.2.3   Repudiating bids
Once an RA has sent its bid it may happen that this RA is awarded the connection in the auction phase. However, there is no real proof either that this bid was sent by the RA, or that the bid was received by the CA.

We can imagine two attacks. First, an RA regrets its bid and now wishes to renege and so refuses to make the connection it has been awarded. Secondly, the CA denies it has received a bid from an RA and gives the connection illegally to another RA. Both, RA and CA, can complain about their situation, but they would need proof.

6.3.3   *Scalability and reliability*

6.3.3.1   Limit of connecting PUAs
One big problem is scalability: how many PUAs can connect to the CA without causing it to collapse under the volume of requests? A simple attack can occur unconsciously when there are many PUAs trying to connect to the same CA at the same time. In a real life implementation of this scheme, the scalability problem must be solved. In fact, in the architecture described in section 6.2 there could be several CAs, even to the extent of being keyed by source and destination.

6.3.3.2   Persistence of agents
The CA is the key element of the system, so it has to be always on-line. If the CA should shut down, all the connections through this source (or source-destination pair) are blocked. Because of this, a possible attack on the network is to try to bring down the host of the CA. Another attack can be done by trying to bring down the host of an RA.

6.4   Security Mechanisms

We have described above some problems of the scheme introduced in section 5.2. We turn now to revising some security mechanisms we can use to solve these problems. All these mechanisms mentioned in this section are further explained in Scheneier (1996.)

6.4.1   Public key infrastructure

Public key cryptography requires the use of two keys. One of these keys is kept private while the other is made public. The technique can be used both to send secret messages and to authenticate a message.

A Public Key Infrastructure (PKI) consists in a set of algorithms to implement public key cryptography, a Trusted Authority (TA) with a key pair, a certificate standard, an information directory to place all users' public keys in the form of TA-signed certificates and mechanisms to access this directory and to apply all cryptographic algorithms.

Usually, a PKI includes the RSA algorithm, X.500 directory, the LDAP protocol to access the directory, X.509v3 for certificates, with the TAs in a hierarchical structure. TA can be organised in hierarchies to adapt to specific trust models. If this is the case, specific trust policies must be defined for clients. We must note that identical hierarchical structures can implement different trust models, since we can define several trust relationships (such as trust only on my TA, or also on all TAs trusted by my TA). Scheneier (1996) describes all these PKI elements.

6.4.2   Non-repudiation protocol

A non-repudiation protocol (Zhou and Gollmann, 1996) is a sequence of messages interchanged between two parties in such a way that it is impossible for either of the parties to deny having sent or received a piece of information. Normally, receipts are used so as to be able to demonstrate sending or receiving a message to a third party.

6.4.3   Secret sharing schemes

A secret sharing scheme is a cryptographic protocol which allows a secret to be shared among, say, n parties. From an original secret, n shares are generated and distributed. Each share is useless on its own. With a collusion of at least k shares, the original secret can be reconstructed. n and k ($k < n$) are parameters of the mathematical secret sharing scheme.

Some secret sharing schemes allow the detection of cheaters (Rifà, 1993), i.e., parties trying to alter the usual secret recovering process.

6.4.4   Tamper-proof environments

In such environments, no specific security strategy is needed due to the assumption of the existence of tamper-proof hardware (as described in Palmer, 1994). Proposals using this sort of environment are necessarily very restrictive, since this special hardware must be available on each host in the system.

6.4.5   Security in mobile agent systems

Mobile agents extend the static distributed systems model. Programs are executed in computers that are not administered by the program owner. Obviously, a guarantee is needed to avoid attacks from the program on the host and from the host on the program (Hohl, 1998a). In our business model one possibility is for the SP to upload its RA, which contains the bidder, into the NP's host. Notice that the business model does not require uploading code from the user.

We know the threat of viruses, worms, trojan horses, etc. We can't know, *a priori*, the nature of arriving programs. Security aspects of mobile agents must be well understood and mechanisms to solve them must be implemented. This is essential in this system, where mobile agents are a technological key. There are some existing mechanisms that offer partial protection but, unfortunately, there is not yet a global solution for agent security. Several aspects remain open and are the subject of research.

There are two main security areas in mobile agent systems: a) security between agents and b) security between agent's hosts. Other areas, such as host to host security or host third parties security, are not considered here. Security aspects of these last areas, such as impersonation, are already included in the former or are well known (Pfleeger, 1997).

a) Agents may be attacked by other malicious agents. These attacks include data and code manipulation, when physical access to data and code areas is possible; impersonation, i.e. an agent trying to act as if it were another one; cheating, if the agent uses a service it has not paid for; and denial of service, for instance filling an agent message buffer. To prevent these attacks some mechanisms exist: safe languages; sandbox security model (resource filtering), as in the Java Sandbox model (Gong,. 1997); or agent authentication using a PKI.

b) A host must be protected against malicious agents and agents against malicious hosts. However, security between agents and hosts is not a symmetrical problem when solution difficulty is considered. The first case (for example, protecting the NP from code uploaded from the SP) is similar to agent to agent security described above. The second case, i.e. protecting agents against hosts, is a hard problem to solve and is discussed in the next subsection.

6.4.5.1   Malicious hosts problem

In mobile agent systems, program operator and host administrator are not the same person. Thus, the possibility of the execution environment (in the host) attacking the agent is a fundamental issue in these systems.

Which actions are considered an attack depends on which guarantees are needed by the agent owner to let it execute. As we see (Robles, 1999) these attacks can be identified:

- Code, data and flow control spying
- Code, data and flow control manipulation
- Wrong code execution
- Host impersonation
- Execution denial
- Agent interaction spying
- Agent interaction manipulation
- False system-call return values
- Host crashing to kill an agent

- Agent re-execution
- Itinerary manipulation

A description of each attack can be found in Robles (1999). Some authors state this problem has no solution (Harrison et al. 1995), but some partial solutions can be found. White (1994) with the Telescript systems, and Cheng (1997) with a security infrastructure are examples of this. Both proposals are based on host infrastructure controlled by one operator. These approaches are closed since the access to the host is supervised.

Different solutions are given in Yee (1997), Strasser et al. (1998) and Baek, (1998) to avoid re-execution of code or data manipulation. They are very specific solutions designed to solve particular problems. Vigna (1998) describes a method to detect and prove whether data modification has occurred *a posteriori*: using what he calls cryptographic traces. A formal method to avoid malicious code manipulation by the host is found in Robles (1999). This is based on the 'Proof-Carrying Code'. This method was first proposed in Necula and Lee (1996) and uses a logical framework to prove some security properties of code. As this is a formal method, it is difficult to implement. Other proposals, such as mobile cryptography (Sander and Tschudin, 1997), are just mathematical models and it is not clear that they can be implemented.

Finally, in Hohl (1998b) a general solution is given to solve the problem: the Black Box. This method consists in obfuscating the code, for example by scrambling code lines, and then giving an appropriate time window in which to execute. It is supposed that the code cannot be de-obfuscated in that time. With a scheme like this, a host can execute a program without knowing it, and attacking it. As discussed in Robles (1999) this is a weak solution: with faster machines and parallel programming this code expiration time needs to be shortened, but there are no clear ways of knowing how this would be done as we do not know the architecture of the host.

6.5   Solving Security Problems in Auctions with an SP

6.5.1   Using a Public Key Infrastructure

From our point of view, a PKI with a Trusted Authority (TA) is needed in the architecture presented above. Using a PKI many problems described above can be solved directly, and it provides a framework to embed the solutions for the other security problems.

With such an infrastructure, and by using standard methods of secure (secret and authenticated) communications (such as the SSL protocol (Freier and Karlton, 1996) or the STS protocol (Diffie et al, 1992)), problems related with authentication and secrecy 5.3.1.1 (authentication PUA↔CA), 6.3.1.2 (authentication RA↔CA), and 6.3.1.3 (insecure channels) would be solved. It is important to point out that this infrastructure would also bring a framework for electronic payment.

Problem 6.3.2.3, namely repudiating bids, can be solved using a non-repudiation protocol between RAs and CA over a PKI. A regulatory body can penalise SPs and CAs based on the receipts generated. A discussion of the operation of a non-repudiation protocol can be found in Riera et al. (1998).

New problems then arise: How may certificates be distributed? How will private keys be protected? Should all the information collected by the CA, over many auctions, be kept secret? Who owns the TA? These questions are discussed in the next sections.

6.5.2   Regulatory Board

An important question to solve is deciding who owns the Trusted Authority (TA). It seems to be logical to give the TA responsibility to a regulatory body. But, in doing this, all trust in the system would fall on this regulatory body. Perhaps, such a TA could be bribed by a powerful Network Provider.

A similar problem occurred in Riera and Borrell (1999) when designing an electronic voting scheme. The solution was to create an Electoral Board using a secret sharing scheme. A description of the Electoral Board mechanisms can be found in Riera et al. (1998). The board system may also be applied to solve the problem of owning the TA. A set of regulatory bodies, call it a Regulatory Board (RB), controls the TA. We can easily imagine an RB made up by a chosen subset of NPs, SPs, user co-operatives and end user organisations. To corrupt the trust scheme, a collusion of some entities needs to occur, but the constitution of the RB is designed to make this difficult to happen (parties with opposed interests are not likely to collude, as in the case of different political parties in an Electoral Board). Trying to cheat such a control system is detectable if using some variant of the secret sharing scheme as described in Rifà (1993). With this model nobody directly controls the TA or, which is the same, the TA is self-owned, and it has a distributed control.

6.5.3   Persistence of agents

Some agents in Bigham et al. (1999a) must be always on-line. This is the true for the CAs and RAs, for instance. If these agents stop executing (e.g. by an accidental crash, an electricity failure, or sabotage) all system operations are halted. Persistence mechanisms have to be provided in order to a) keep all agents working properly all the time, and b) recover the state of the system before the crash. There are schemes that solve this problem, as seen in section 6.4.5, like the proposal described in Strasser et al. (1998). Thus, we could avoid problem 6.3.3.2.

Persistence of RAs could perhaps be achieved using a different scheme. A new bidding protocol could replace the actual protocol used in the system. The augmented protocol would monitor RA activity so that a breakdown would be detected. As we will see in section 6.5.5, the design of this protocol is an open problem that has to be solved.

6.5.4   Protecting the CA physically

Since the CA is the key element of the architecture, it is also the main target of both network providers and users' attacks. Some extra protection can be given to hardware executing these agents. CAs could execute under tamper proof environments (TPE) to avoid attacks against its execution. A proposal of a TPE can be found in Palmer (1994) with an introduction of a secure crypto-processor.

6.5.5   Bidding protocol

We need an improved bidding protocol to solve problems 6.3.2.1 (same opportunity to all bidders) and 6.3.2.2 (preventing denial of service attacks) in the auction phase of the scheme. This new protocol could also solve the bid repudiation problem (6.3.2.3).

This protocol must guarantee equal opportunities to all auction participants, with independence from its distance and access type. Also, it must prevent denial of service attacks. To achieve this, it is not enough to use a simple one-way stateless protocol as in Bigham et al. (1999b) As we have seen in section 6.5.1, it is possible to solve the bid repudiation problem by using a non-repudiation protocol in a PKI. If the auction protocol is enhanced with the non-repudiation abilities of Riera et al. (1998) there will result a complete protocol solving both problems. This protocol can be also used to monitor RA activity detecting agent failures, i.e. could provide persistence of agents (see section 6.5.3).

A protocol like this requires a meticulous design. Actually, this is a complex open problem that has to be solved. This protocol could be used in similar applications related to auctioning. In the design of the auction in Bigham et al. (1999a,b) the RAs for a source-to-destination pair are placed in the source node, alleviating the problem. It is up to the bidders, using SP proprietary code, to make a bid before time-out. (Note that the source is where the connection is initiated. Sometimes the bulk of the bandwidth usage can be in the opposite direction.) However, if the SP's bidders were in different machines, i.e. not hosted by the NP, then this problem is important since each SP bidder would be in a machine owned by this SP.

6.5.6   Key distribution

Once using a PKI, it has to be decided how certificates are to be distributed. Secure mechanisms exist to access a directory. Two examples are X.500, in the ISO OSI standard and LDAP over TCP (Scheneier, 1996).

Another problem is how to protect private keys. As users must authenticate, they need to carry their private keys. Special devices exist to carry private information, such as smartcards. What happens if some non-human party in the system needs to have private information, as in the case of the CA or RA? (We can imagine the auctions extending to payment.) We cannot rely on wearable devices to protect against theft. To solve a similar problem in a voting scheme, we used a smartcard to protect the private key of non-human elements (Electoral Colleges) (Riera et al. 1998). Since this smartcard could be stolen, we needed special physical security on these elements (such as physical locks and keys). Unfortunately, using such security measures results in the loss of mobility. Mobile agents cannot carry the smartcard when they travel on the net. Using such physical devices involves other drawbacks. Smartcards need to be personalised with owner information. This job has to be done by the Trusted Authority, since it is the only party in the system knowing private keys.

6.5.7   Scalability

To solve problem 6.3.3, i.e. scalability, we need to extend the model where there is just one CA allowing user connections to the network. If we add a new agent, say a Request Distribution Agent (RDA), whose mission is redirect user (PUA) connection requests to any of the CAs available, we have solved the problem by decentralising the activity of the CA into many CAs.

To implement this RDA scheme, our system has to fulfil some requirements. First, all RAs must have access to all CAs. This can be achieved by locating CAs and RAs on the same logical network. We note that this network can be different from that used by RDAs to access CAs. In the context of the architecture Bigham et al. (1999), it is easily possible to create a CA for each source destination pair. Whether this is necessary depends on the typical loads at each source and the loads to each source destination pair. An RDA scheme could exploit this. Finer granularity of protection could be achieved by partitioning (randomly) the load for every particular source destination pair.

6.5.8   Others Problems

After many auctions a CA will have collected a lot of information about users and Service Providers. With all this information both user and SP profiles can be computed. Mining of this data gives valuable information which could be used against the user or SP.

Thus, the CA code must fulfil some requirements in order to protect privacy. For instance, information collected during an auction has to be destroyed when the auction is finished.

It is the job of the Regulatory Board to define what constitutes safe code behaviour. It could do this by specifying a *safety policy*. Some mechanisms have to be provided in order to guarantee that the CA carries this safety policy out. A logical framework can be used to achieve this, as described in Necula and Lee (1996). This scheme could also be extended to the RAs, e.g. over allocation of spare capacity.

6.6   Conclusions

In this chapter we have analysed many different kinds of security deficiencies in Bigham et al. (1999). We have sorted them into three classes: authentication and security, bidding protocol, and scalability and reliability. After a brief description of some of the important mechanisms used to solve generic security problems, we have given some solutions to particular security problems associated with auctioning bandwidth within the architecture described. These solutions are based on techniques used to solve security problems in electronic voting schemes.

A Public Key Infrastructure (PKI) has been used to solve some deficiencies. Over this PKI we have suggested a method to avoid regulatory body corruption: the Regulatory Board (RB). This RB is based on the voting system Electoral Board and uses a secret sharing scheme to allow a distributed control and ownership of the Trusted Authority. Some methods have been pointed out to solve such mobile-agent-specific problems as agent persistence or physical protection. To solve system scalability we have introduced a Request Distribution Architecture, which decentralises Connection Agent activity.

Solutions described in section 6.5 solve or point out how to solve deficiencies set out in section 6.3. Nevertheless, some of the solutions introduced are just sketched, such as the Regulatory Board, the enhanced bidding protocol, and the RDA scheme. Since they are open problems they will be developed in future work.

Problems still arise using all these methods. These include problems related to distributing system elements over a network, certificate distribution, private key protection and real-time service. In future work we also plan to investigate further the constraints imposed by real-time requirements.

Acknowledgements

Contributions of S. Robles and J. Borrell have been partially funded by the Spanish Government Commission CICYT, through its grant TEL97-0663.

6.7   References

Baek, J. (1998) A Design of a Protocol for Detecting a Mobile Agent Clone and its Correctness Proof Using Coloured Petri Nets. Technical Report TR-DIC-CSL-1998-002, Kwang-Ju Institute of Science and Technology.

Bigham, J., Cuthbert, L. G., Hayzelden, A. L. G. and Luo, Z. (1999a) Flexible Decentralised Control of Connection Admission. In *Proceedings of IMPACT 99*, pp. 5-15, Seattle WA, ISBN 0904188647

Bigham, J., Cuthbert, L. G., Hayzelden, A. L. G. and Luo, Z. (1999b) Agent Interaction for Network Resource Management. *ICON - Interoperable Communication Networks Journal*. Vol. **2(1)**. Published by Baltzer Science Publishers, Special Issue on Network and Service Management in the Broadband Area.

Borrell, J. and Rifà, J. (1996) An Implementable Secure Voting Scheme. *Computers & Security*, **15(3)**: 327-338.

Cheng, Y. (1997) A Comprehensive Security Infrastructure for Mobile Agents. Master thesis, Stockholm University/KTH.

Diffie, W., Van Oorschot, P. C. and Wiener. M.J. (1992) Authentication and Authenticated Key Exchanges. *Designs, Codes and Cryptography*, **2**: 107-125.

Freier, A.O. and Karlton, P. (1996) The SSL Protocol, Version 3.0. Netscape Communications Corporation.

Gong, L. (1997) Java Security Architecture for JDK 1.2. Sun Microsystems, Inc. Montain View.

Harrison, C., Chess, D. and Kershenbaum, A. (1995) Mobile Agents: Are They a Good Idea? Technical report, IBM, IBM T. J. Watson Research Center.

Hohl, F. (1998a) A Model of Attacks of Malicious Hosts against Mobile Agents. In *Proceedings of the ECOOP Workshop on Distributed Object Security and 4th Workshop on Mobile Object Systems*, pp. 105-120.

Hohl, F. (1998b) Time Limited Blackbox Security: Protecting Mobile Agents from Malicious Hosts. *In Mobile Agents and Security*, LNCS **1419**. Springer-Verlag, pp. 92-113.

Necula, G. C. and Lee, P. (1996) Proof-Carrying Code. Technical Report, School of Computer Science, Carnegie Mellon University.

Palmer, E. (1994) An Introduction to CITADEL - a secure crypto coprocessor for workstations. In *Proceedings of the IFIC-SEC '94*.

Pfleeger, C. P. (1997) *Security in Computing*. Prentice Hall International Editions.

Quian, T. (1998) Cherubim Agent Based Dynamic Security Architecture. Technical report, University of Illinois at Urbana-Champaign.

Riera, A. (1999) Design of Implementable solution for Large Scale Voting Schemes. PhD thesis. Universitat Autonoma de Barcelona, Bellaterra.

Riera, A. and Borrell, J. (1999) Practical Approach to Anonymity in Large Scale Electronic Voting Schemes. In *Network and Distributed System Security, NDSS '99*, pp.69-82.

Riera, A., Borrell, J. and Rifà, J. (1998) An Uncoercible Verifiable Electronic Voting Protocol. In *IFIP-SEC '98*, pp. 206-215. Austrian Computer Society.

Rifà, J. (1993). How to Avoid the Cheaters Succeed in the Key Sharing Scheme. *Designs, Codes and Cryptography*, **3**:221-228.

Robles, S. (1999) Applying Mobile Agent Systems in the Design of a Large Scale Voting Scheme (in Catalan). *Master thesis*. Universitat Autonoma de Barcelona, Bellaterra.

Robles, S., Pons, J. and Borrell, J. (1998) Design of a Secure Large Scale Voting Scheme: The Electoral College (in Spanish). In *Proceedings of the V Spanish Meeting of Cryptography and Information Security*, pp. 333-344, Universidad de Malaga, Malaga.

Sander, T. and Tschudin, C. (1997) Towards Mobile Cryptography, Technical Report 97-049, International Computer Science Institute, Berkely, www.icsi.berkeleyedu/~sanders/publications/tr-97-049.ps

Scheneier, B. (1996) *Applied Cryptography. Protocols, Algorithms, and Source Code in C*. John Wiley & Sons, New York.

Strasser, M., Rothermel, K. and Maiofer, C. (1998) Providing Reliable Agents for Electronic Commerce. In *Proceedings of the International IFIP-GI Working Conference*, Hamburg, Germany, LNCS 1402. Springer-Verlag, pp. 241-253.

Vigna, G. (1998) Cryptographic traces. In *Mobile Agents and Security*, LNCS **1419**. Springer-Verlag, pp. 92-113.

White, J. E. (1994) Telescript Technology: The Foundation for the Electronic Marketplace. White Paper, General Magic Inc.

Yee, B. (1997) A Sanctuary for Mobile Agents. In DARPA Workshop on Foundations for Secure Mobile Code Workshop. URL: http://www.cs.nps.navy.mil/research/languages/statements/bsy.ps

Zhou, J. and Gollmann, D. (1996) Observations on Non-Repudiation. In *Asiacrypt '96*, LNCS **1163**, pp. 133-144.

7

Secure Payments within an Agent-Based Personal Travel Market

R. Kerkdijk

7.1   Introduction

Agent technology is beginning to be used to develop real solutions to real business problems. However, most present-day uses of agent technology are 'single-supplier' solutions that is, the agents involved have a strong family relationship, having been developed within a well-defined engineering community to meet specific objectives. Any communication between agents is conducted using proprietary protocols developed to meet the specific needs of the application. In the future, however, agents developed independently by different companies should be able to communicate reliably and effectively. Single-supplier solutions will not be acceptable. Standards will therefore be required that ensure 'open' interaction between heterogeneous communities of agents. The Foundation for Intelligent Physical Agents (FIPA) was formed in December 1995 with the aim of developing such standards, the first versions of which were released in 1997.

The goal of the FACTS (FIPA Agent Communication Technologies and Services) project, running for two years (February 1998 to February 2000), is to validate the work of FIPA by constructing a number of demonstrator systems based on proposed standards. The project will take a 'do and discover' approach to standards development. Work Package 3 of the FACTS Project deals with agent-based Electronic Trading, in particular the development of a Personal Travel Market (PTM) using FIPA-compliant agent technologies and the Internet.

Secure payments form an essential part of such an agent-based travel market, or any E-Commerce service for that matter. Within FACTS WP3, the need for a secure payment mechanism has been acknowledged. This chapter will elaborate upon the work that has been done so far to incorporate such a mechanism within the PTM infrastructure.

7.2   An Agent-Based Personal Travel Market

The combination of the Internet and software agent technologies will radically change electronic commerce. The Internet opens up electronic services to anyone with a computer and a modem, and provides effective market access for many service providers. With software agents, responsibility for purchasing or selling goods and services can be delegated to a piece of autonomous software, i.e. an agent. Travel retailing is a good example of an industry that can be revolutionised by these technologies, and is used in FACTS WP3 to validate the FIPA standards.

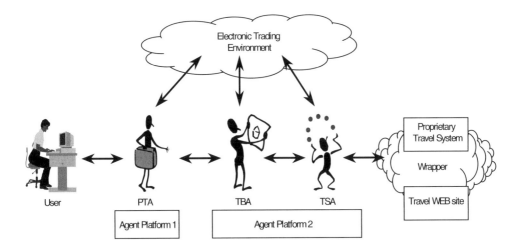

Figure 7.1 PTM Reference Model

Within the PTM application three types of trading agents are incorporated, corresponding to the identified types of parties involved in such a market place (see figure 7.1).

7.2.1   Personal Travel Agents (PTAs)

The Personal Travel Agent (PTA) is a personal agent acting as the interface between the user and the Personal Travel Application. Thus it represents the interests of the user within the system. It is responsible for fulfilling the user's travel requests through interactions with Travel Broker Agents. The PTA interacts with Travel Broker Agents in order to find travel plans that satisfy both the user's explicit travel request as well as the user's personal preferences. The PTA is also responsible for protecting the privacy of the user's preference profile while interacting with Travel Broker Agents in order to augment travel plans with 'soft' or 'value-added' services. This means offering suggestions and/or making bookings for services that may be attractive to the user on the basis of their personal interests, for example theatre or football. Furthermore, the PTA will be integrated with a secure electronic payment system.

7.2.2   Travel Broker Agents (TBAs)

The role of a Travel Broker Agent (TBA) is to act as an intermediary between the Personal Travel Agent and Travel Service Agents. In addition to this, the TBA can use its position to offer value-added services to the user. Typically, it will collect a trip request from the PTA, analyse it, identify the different segments and try to compose a trip plan that satisfies the user's requirements, based on schedule and pricing information provided by the TSAs. This agent may have a company profile expressing the preferences of the user's company in business travel. This profile will be used in connection with the user's profile in order to direct the trip search process.

7.2.3   Travel Service Agents (TSAs)

The Travel Service Agents are responsible for providing data and services to the Travel Broker Agent. In order to schedule and propose travel plans to the user, the TBAs will need to access a wide range of travel information such as airline tickets sales, railway timetables, hotel room availability and prices, weather reports and so on.

Each agent resides within an agent platform, which provides the physical infrastructure in which agents may be deployed. This is also illustrated in Figure 7.1. Furthermore, platforms provide a number of common agent services such as directory facilities and naming services. In addition to this, facilities such as secure payments may be integrated within an agent platform in order to promote re-use.

Note that Agent Platform 1, with the PTA residing in it, may be located at the user premises (i.e. on their personal PC), whilst Agent Platform 2 may be regarded as the Travel Retailing environment for companies on the Internet. Although definite choices concerning this issue remain to be made, this presents a likely alternative. However, one should take into account the size of the software that needs to be installed by the user for this purpose. If this becomes very large, users will find the application unattractive.

7.3   Secure Payments

Agents may be authorised to pay for the travel services offered to them. For this purpose, some form of secure payment mechanism needs to be incorporated within the infrastructure. The following requirements have been defined regarding such a mechanism:

* Payments performed by agents should be based on some form of standard payment protocol
* The applied protocol should offer a high level of security
* For the travel application, creditcard transactions seem to be the best solution

Based on these constraints, two alternative protocols may be applied, namely SSL (Secure Sockets Layer) and SET (Secure Electronic Transaction). Both will be elaborated upon in the following sections.

7.3.1   SET

The SET (Secure Electronic Transaction) protocol is an upcoming industry standard for performing secure creditcard transactions over the Internet. Visa, MasterCard and various other companies[1] were involved in its development, which was largely motivated by concerns about possible growth in fraud, due to unsecured creditcard transactions on the Internet. SET provides:

- Confidentiality of both payment information (i.e. creditcard number) and order information
- Payment integrity
- Authentication of both merchants and cardholders.

One of the most attractive features of SET is that the creditcard number is at no point sent to the merchant in clear text. Figure 7.2 is the SET Mark logo.

Figure 7.2 The SET Mark[2]

In figure 7.3, the standard infrastructure for performing SET-based payments is shown. Most entities depicted here are straightforward analogies to real-world creditcard payments. A few, however, need further explanation. These are described below.

One of the main issues when providing secure payments is authentication of the involved entities. SET uses a robust set of digital certificates for this purpose. These certificates are an essential part of SET's Public Key Infrastructure (PKI). Each participant in a SET transaction requires a specific certificate or set of certificates that not only uniquely identifies this participant, but also attests to his/her privilege as holder of a payment card or his/her privilege as holder of a Merchant account. Brand Associations (Visa/MasterCard) or Card Issuers commission so-called *Certificate Authorities (CAs)* to

---

[1] Among others: Netscape, Microsoft, IBM, RSA and Verisign.

[2] The SET Logo or SET Mark is a visible symbol signifying that software complies with the SET Specification.

carry out the work of managing SET digital certificates. The exchange of certificates enables entities to check digital signatures, thereby verifying the source of messages. Furthermore, it enables these entities to encrypt messages such that the content may only be read by the intended recipient. That is, information is exchanged on a 'need to know' basis only, so that financial data is not available to the merchant and product data is not available to the acquirer.

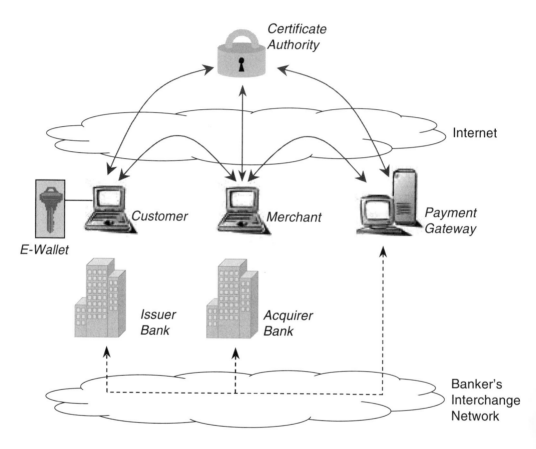

Figure 7.3 SET infrastructure

Complementary to this, SET introduces the notion of a *Payment Gateway*, which serves as a link towards the Banker's Interchange Network. There are two main reasons for establishing a separate entity, different from the Merchant Server, for this purpose:

- Private information, such as creditcard numbers, need not be communicated to the merchant in this setup, thereby establishing a higher level of security (confidentiality).
- The Payment Gateway preprocesses authorisation, capture and settlement work, which provides a high level of efficiency.

Another fundamental requirement for performing SET payments is a component which enables secure handling of private information, such as private cryptographic keys, on the user side. In most implementations this component is known as the *Electronic Wallet (E-Wallet)*. This wallet embodies the SET protocol on the customer side and provide a means to store and manage the certificates to digitally sign messages, along with the security aspects consumers demand to keep private data private. Examples of E-Wallets are the GlobeSet Wallet (Globeset, 1999), the Microsoft Wallet (MSWal, 1999), IBM's CommercePOINT Wallet (IBMCom, 1999) and Trintech's Paypurse (Trint, 2000).

More detailed information on SET may be obtain from SET (1997), SETCo (1999) and Merkow (1998).

7.3.2   SSL

Netscape has developed a protocol called Secure Sockets Layer, which provides transport-level security for TCP/IP connections. This protocol has, in a slightly different form, been issued as a standard by the IETF (Internet Engineering Task Force) under the name TLS (Transport Layer Security). Using SSL (TLS), which is embedded in both Microsoft's and Netscape's browsers, all communication between a certain client and server may be encrypted.

A minimum requirement for creating an SSL/TLS connection is that the server needs to possess a digital certificate, which may be used to authenticate it. To establish mutual authentication, the client needs to possess such a certificate as well. Once the authentication phase has been performed successfully, client and server negotiate a session key, which is then used to encrypt all communication. As was the case with SET, certificates are issued by Certificate Authorities.

Within the PTM context, SSL can provide:

- Confidentiality of exchanged messages
- Message integrity
- Authentication of both merchants and cardholders

SSL is at present widely used to secure creditcard payments over the Internet. More information on SSL may be found in Netscape (2000)

7.3.3   Direction chosen

When comparing the two alternatives described above, the following issues arise:

- The SSL/TLS scenario implies that the merchant receives the cardholder's credit card number. The merchant may then re-use this number in a malicious fashion, or store it in an insecure database. The cardholder will therefore have to trust the merchant to treat creditcard numbers with care and integrity. However, it is impossible to reinforce this technically. Within SET, however, the merchant does not acquire knowledge of the credit card number.

- Within SET, a direct link with the banker's interchange network is established, which is attractive from an efficiency point of view. SSL/TLS, however, does not define such a link. Therefore, to translate the SSL/TLS-based payment into actual credits and debits, a separate procedure needs to be applied.
- SET provides application-level security, whilst SSL/TLS is performed at the transport level. As a consequence, SSL/TLS can only establish that the connection between merchant and cardholder is tamperproof. It can by no means ensure that the cardholder will in fact receive the actual order. SET on the other hand signs and encrypts its messages at the application level, thereby enabling the cardholder to receive a signed confirmation of the order placed. Such a signed confirmation may be of great value in case of disputes. Furthermore, the presence of the Payment Gateway adds an extra check on the validity of the merchant, which is not defined within SSL/TLS.

Based on these considerations, it is concluded that from a technical point of view SET is the more attractive tool to enable agents to perform creditcard payments within the PTM, since it offers a level of security that is higher than that of SSL/TLS. This project will therefore focus on incorporating SET-based creditcard payments within the PTM infrastructure.

It should be noted that implementing SET is not a straightforward task. That is, obtaining and integrating SET software components may entail a substantial effort. Specifically the Payment Gateway is a very complex and expensive piece of software. Therefore, the possibility should not be ruled out that somewhere along the lines of the project an alternative solution should be pursued. This possibility will be taken into account in the architectural design.

7.4   Architectural Integration

Within the PTM, we would like to delegate the task of performing SET credit card transactions to agents. For the purpose of this trial, the following constraints have been defined:

- Obtaining certificates is not a task that users will want to delegate to their agents. Furthermore, banks and CAs will require the user to be involved in this process for identification and authentication purposes. Therefore, we assume all certificates and the E-Wallet to be in place.
- Payments between companies are in general not performed by means of creditcard. Therefore, SET payments shall only be performed between the PTA and the TBA, business-to-business payments are seen to be outside of the scope of the PTM
- As much as possible of the standard SET infrastructure shall be kept intact. Thereby the inherent security of SET payments shall remain present and the necessary alterations when implementing shall be limited.

Based on these constraints, the infrastructure shown in Figure 7.4 has been designed.

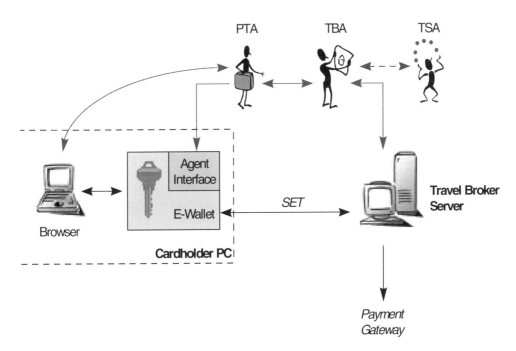

Figure 7.4 SET within the PTM

The PTA interfaces both with the browser on the cardholder's PC, in order to communicate with its owner, and with the E-Wallet, in order to initialise payments. For this purpose, the PTA will need to send a 'wake-up message' to the E-Wallet. This message is not formally defined within the SET specifications, but is of vital importance to the payment process. In order to establish communication between the PTA and the E-Wallet, a wrapper-like *Agent Interface* is necessary. The actual SET payment process is performed between the E-Wallet and the Travel Broker Server, not at the agent level. Therefore, during actual payment interaction the level of trust is the same as in standard SET payment.

The PTA will have to be authorised to initialise the E-Wallet for payments. In most implementations of SET the user is prompted to enter their E-Wallet password for this purpose. There are, however, alternative solutions, such as biometric authentication or the insertion of an adequately personalised smartcard.

The PTA and the Agent Interface will have to be implemented such that one of two scenarios may be performed: either the PTA has authorisation to initiate the use of the secret key itself, or, after PTA initialisation, the user is prompted to provide the E-Wallet password. In the latter case, user interaction is necessary. This is not desirable from a usability point of view, but is necessary to keep the content of the E-Wallet secure. Furthermore, it will give users a sense of control over the payment.

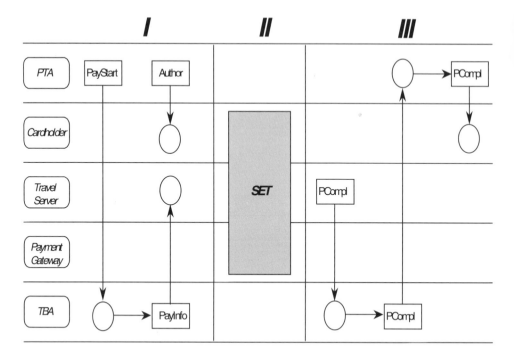

Figure 7.5 Communication procedure

For authentication and authorisation purposes, the PTA will carry a *token*, in which an authorisation code for opening up the E-Wallet is encapsulated.

In Figure 7.5 the communication procedure is shown. The following communication steps are performed:

- In step I, the PTA requests the TBA to pay by creditcard. The TBA then informs the Travel Broker Server of the payment at hand, while parallel to that the PTA initialises the E-Wallet.
- In step II, the standard SET procedure is performed by the E-Wallet, the Travel Broker Server and the Payment Gateway.
- Finally, in step III, after completion of the payment, the merchant informs the TBA of this fact. The TBA passes this message on to the PTA, which then notifies its owner of payment completion.

The infrastructure and message flows are a natural extension of the basic PTM. Therefore, integration of the secure payment mechanism within the PTM may be performed in a straightforward manner.

It should be noted that the infrastructure depicted in section 7.4 could, in fact, also be used to support payments based on protocols other than SET. Most electronic payment protocols require cardholder and merchant software components. Since the actual payment is performed outside the FIPA world, the SET components could be replaced with others without requiring substantial interface modifications. One could, for example, replace the E-Wallet and the Payment Server with specific software components that support the SSL protocol. The payment may then be conducted by sending creditcard information over an SSL connection. Therefore, the architecture shown in Figure 7.4 may be regarded as generic.

7.5   Securing Communication

Since FIPA has to date not specified agent security adequately, securing agent-agent communication is outside the scope of this document. However, the payment process also requires secure information exchange between agents and their owners or back-end systems:

- Communication between the TBA and the Travel Broker Server should be secured, since consumers will demand that their travel details be kept private.
- Communication between the PTA and the cardholder may have to be secured, depending on the location of the PTA. If it resides on the cardholder's PC, security requirements on both storing the token within the PTA and communicating it to the E-Wallet, as well as on informative messages to the cardholder are not as strict as if the PTA resides on a remote platform. The same goes for the level at which the authorisation token needs to be secured within the PTA.

For securing communication, a PKI (Public Key Infrastructure)-based solution seems to be the best alternative. There are two main reasons for this:

- Since PTM-type applications will eventually operate on the Internet, the number of parties involved is potentially large. For secure communication within large communities, key exchange is a problem when applying a secret key scheme. PKI seems to be the best solution for this problem.
- Within an insecure environment such as the Internet, authentication, integrity and confidentiality are of utmost importance. These are all offered by a PKI-based solution.

Therefore, in those cases where secure communication is required, all communicating parties need to possess a certificate specific for this purpose, i.e. other than the SET certificate. Once such certificates are in place, secure communication will be performed as follows:[3]

- Messages will be communicated in encrypted form, using a confidential session key. A symmetric encryption scheme, such as DES, will be applied here.

---

[3] Note that similar mechanisms may be applied to secure agent to agent communication. FIPA is currently developing agent security specifications based on these principles, see FIPA (1998).

- This random key will be sent over in encrypted form, using the public key of the recipient.
- Each message will be signed with the sender's private key.
- A time stamp will be added to the message, in order to prevent replay by malicious parties.

If the PTA resides on a remote platform, the authorisation token will be secured within it in encrypted form, using a random key. A symmetric encryption scheme, such as DES, shall be applied here. This random key was generated on the cardholder's PC for that specific purchase. A new key will be generated for each item that is to be bought by the agent.

It should be noted that the described mechanisms for securing messages require an agent to carry around a private key. Malicious parties (e.g. other agents or the platform in which the agent resides) may attempt (and succeed) to compromise such keys, thereby gaining access to all sorts of private information and systems! At present, no solution has been proposed to overcome this problem.

7.6   Status Quo

At the time of writing, a phase-1 demonstrator has been successfully delivered to the project. Within this demonstrator, all agents and their basic interactions related to negotiating travel services have been implemented. Steps are now being undertaken to integrate secure payments within the existing PTM infrastructure. It is felt that this will add substantial value to the demonstrator as well as to the project as a whole.

As was explained above, the initial plan within the project was to implement secure payments based on the SET protocol, since this option is technologically more attractive than SSL payments. Unfortunately, however, this option turned out to be infeasible. Commercial SET components, as supplied by various software vendors, are very complex and require a substantial implementation effort. More specifically, installing a Payment Gateway is far from straightforward. Since the goal of the project is to implement agent technology rather than the SET protocol, this option was abandoned.

The alternative solution that was chosen is to implement creditcard payments secured by the SSL protocol. Although this is not the most appealing solution from a technical point of view, SSL is still the most widely used protocol to support creditcard payments on the Internet today. The architecture described above (see Figure 7.4) has been kept intact. That is, a Java based E-Wallet and Payment Server will be implemented, which may communicate over SSL. As was explained in section 7.4, the interface towards the PTM will remain practically the same. Therefore the same point is proven.

The final demonstrator, in which secure payments may be performed by agents, was delivered in February 2000.

7.8   Conclusion

Based on what has been described above, we may conclude that agent-supported secure payments, based on protocols like SET or SSL, are feasible. Furthermore, these do not require many substantial alterations to the standard payment components nor to agent infrastructures such as the PTM.

Acknowledgements

The work described in this chapter was performed within the FACTS (AC317) project. Some material has been supplied by the other partners within work package 3 of FACTS: British Telecommunications, Broadcom Eireann Research, France Telecom and ONERA/CERT. This chapter would not exist in its current form without their kind permission. The views expressed in this chapter are those of the author, and not necessarily those of the other FACTS project partners.

7.9   References

FIPA (1998) Foundation for Intelligent Physical Agents, FIPA 98 Specification Part **10**, Version 1.0: Agent Security Management, October 1998

Globeset (1999) http://www.globeset.com

IBMCom (1999) http://www.internet.ibm.com/secureway/commercepoint.html

Merkow, M. S. (1998) *Building SET Applications for Secure Transactions*, Wiley, 1998.

MSWal (1999) http://www.microsoft.com/siteserver/commerce/default.asp

Netscape (2000) http://home.netscape.com/eng/ssl3/

SET (1997) SET Forum, SET Secure Electronic Transaction Specification Book 1: Business Description, May 1997

SETCo (1999) http://www.setco.org

Trint (2000) http:// www.trintech.com

8

Multi-Agent Solution for Virtual Home Environment

N. Fujino, Y. Matsuda, T. Nishigaya, I. Iida

8.1 Introduction

Recently, the mobile computing industry has drawn wide attention with the rapid development of wireless communications and portable devices. This will be accelerated by the deployment of third-generation mobile communication systems, such as IMT-2000 (International Mobile Telecommunication system 2000) or UMTS (Universal Mobile Telecommunication System). IMT-2000 accommodates up to 2 Mbps data per channel and assumes end-to-end IP (Internet Protocol) communications.

One of the biggest issues what kind of service and software will be provided on this infrastructure. The virtual home environment (VHE) service is one example in the third generation of mobile communication, where the seamless communication is provided for mobile users according to such conditions as using a terminal, time, place and situation.

8.2 VHE Concept

Mobile users access network services from different types of terminal such as a desk-top PC, a lap-top PC, PDA, or cellular phone, dependent on the situation and they change the location frequently from wire-line to wireless network. Also visited networks may be a high-speed LAN or a low-quality cellular network.

Mobile users hope to use a printer and personalised service such as mail and news on a visited network as if he/she is in the home environment. So it is important to provide the mechanism so that they can seamlessly access the home network environment of each user from anywhere. This kind of service concept is called Virtual Home Environment (VHE) (Hartmann, Gorg and Farjami, 1998). The system configuration is illustrated in Figure 8.1.

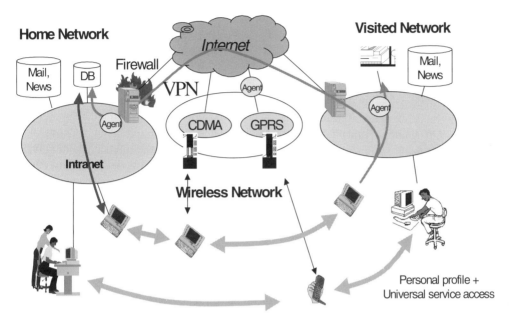

Figure 8.1 Virtual Home Environment

8.3   Requirements for Supporting VHE

The requirements for VHE service are summarised as follows.

8.3.1   Customisation for each user

A user wants to have a service in the same environment at a different location in that user's view. So a user should be able to customise his/her own environment. This is an important point so that telecommunications carriers can distinguish their service from their competitors.

8.3.2   Personal mobility

In general, the relation between user and terminal must be managed to assure personal mobility. A user ordinarily changes his/her terminal dependent on the location; desk-top PC in the office or lap-top PC or PDA outside the office. So personal mobility corresponding to the location should be supported in the next generation.

8.3.3   Service provisioning corresponding to the user profile

In response to the user attributes such as user preference, location, network and terminal, a network must provide the optimum communication methods and convert service contents dynamically.

8.3.4   Off-line operation

In a mobile environment the user is not always connected to the network but in most cases he/she is working off-line, separated from the network. So it is desirable that a user can do off-line operation without worrying about whether the terminal is connected to the network or not. The most important function to do this is data synchronisation.

8.3.5   Automatic network configuration

When a user accesses the network, it is required that the terminal checks the available network and configures the communication environment adaptively. In short, automatic network configuration is desirable.

8.3.6   Automatic service configuration

When a user connects a terminal to a visited network, the mechanism to know what kind of service is available in the network is needed so that the user can use the local printer or mail service at once.

8.3.7   Compensation of network quality

Each network has a different quality. Especially in radio network, the problems relate to unstable channel and high error rate (Imielinski and Korth, 1996) should be hidden from applications or users.

8.4   Multi-Agent Solution

In the VHE service, each service function must work independently or collaboratively corresponding to user situation, and the service function must dynamically be modified with the user requirements. For that purpose, we apply the multi-agent architecture where independent function can interact with the common communication interface as an agent, and a new service function can be added or removed dynamically.

The requirements (1) and (2) can be solved by the personal agent (PA). Since the VHE is a kind of personal communication service, the object representing the user view (PA) is needed (Fujino, Tokuyo and Iida, 1998). Requirement (3) is realised by the collaboration between PA and upper-layer services.

Requirements (4), (5) and (6) are mainly realised by network agents (NA) on both client and server sides. Radio agent (RA), a part of NA, solves the requirement (7) in the case when a user communicates with wireless link.

8.4.1   Personal agent

A personal agent (PA) is assigned for each end user and provides the user with customised communication services and personal view for the network (Iida, Nishigaya and Murakami, 1995).

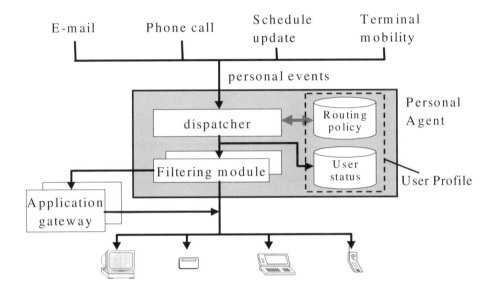

Figure 8.2 Structure of personal agent

Figure 8.2 shows the structure of a PA, which has the following two functions: one is a user profile and the other is an event dispatcher. User profile manages personal attributes such as user preference, terminal profile, communication mode, current location and so on, which enable its user to customise services. Location information is acquired together with area agents (AAs).

A dispatcher executes related tasks according to the routing policy when it receives a communication event such as e-mail transfer or connection request. PA always traces its user location to decide the optimum communication method, and guarantees the personal mobility of each user.

PA basically resides in a network server but can move within the network along with the movement of the user. PA can be created in the terminal as a clone to realise offline operation.

8.4.2   Network agent

The network agent (NA) virtualises communication media and hides the difference of the network from users and services. Figure 8.3 illustrates the architecture.

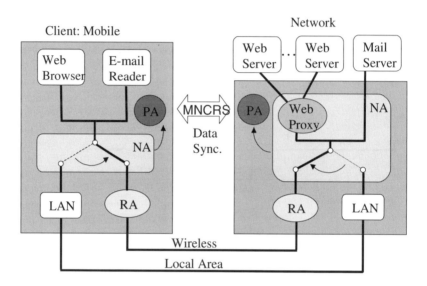

Figure 8.3 Architecture of Network Agent

Network agents are installed on both client and server sides. NA on the client resides in a terminal to check available communication channel such as LAN, telephone, wireless or IR channel. It selects the optimum communication medium and sets up the network environment automatically in accordance with the routing policy managed by PA. This enables its user to guarantee the proper operation and dynamic module addition. NA on the server side resides in the subnet and tells the NA on the terminal side available resources such as printer, mail server, news server and web proxy. NA on the clientside collaborates with that on the server to set up the network configuration automatically. In this way, a user can use the services in a visited network unconsciously.

In order to support offline operation, NA on the client supports data synchronisation to guarantee the consistency of the state of agents on the client. This is realised as a synchroniser defined by MNCRS (Mobile Network Computing Reference Standards). Offline computing function is realised by the mechanism that PA on the server and PA (clone of PA) on the client are synchronised by the above function.

Every change in the master PA that occurs while the terminal keeps offline is automatically synchronised when the terminal is connected to the network again. NA on the client can refer to the customised routing policy through the replicated PA on the client, even if the terminal is offline.

8.4.3   *Radio agent*

Figure 8.4 shows the architecture of a radio agent (RA). RAs are placed on both server and client sides. RAs on the two sides cooperate to hide the low quality of the radio channel, and attain the efficient use of network bandwidth assigned to a user. This function is realised by terminating the TCP/IP and applying a light-weight protocol supporting automatic reconnection and local link control for reducing the connection time. In addition, RA has an interface to tell the effective transmission rate of radio channel to realise the data compression by an adaptive image conversion.

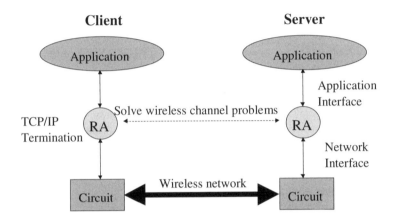

Figure 8.4 Architecture of Radio Agent

The implementation of RA has two types. One is to implement RA as an application proxy and the other is to implement as a network driver. The latter method can carry the IP packet transparently to the terminal. Figure 8.4 is the RA implementation as a network driver combined with NA.

8.5   Agent Platform

We realise PA, NA and other characterised agents using a Java-based agent library called Pathwalker (Ushijima et al. 1998). In this section, we introduce the particular features of Pathwalker.

8.5.1   *Pathwalker*

Pathwalker is a multi-agent platform based on asynchronous message communication. Its main features are as follows:

- Simple Agent Action model
- Dynamic modification of agent by replacing the action
- Data flow model using message pattern matching

In the Agent Action model, all agents in a system are completely the same, but each agent is characterised by having the different action implementations that are the definition of specific behaviour of agents. Since an agent is allowed to add, remove or replace any action implementation at any time, it can change its own functionality and responsibility in the runtime environment. This means that agents are inherently highly independent components, but by injecting the actions as some interaction knowledge into them, we can make them work collaboratively.

Another important feature of Pathwalker is message pattern matching. Every action object declares the expected input message pattern. When an agent receives a message, the agent compares the received message to all of the input pattern of its own actions and selects the appropriate actions. This flexible action selection helps to realise dynamic decision of data flow represented by action chaining, which is applicable to PA's service routing function using the most suitable communication device.

8.6   Prototype System

We developed the VHE prototype system using this architecture. This system has the following features:

- Automatic selection of optimum communication method for user's environment
- Network plug and play without annoying network configuration
- Off-line operation of mobile terminal

Figure 8.5 shows the configuration of this prototype system. Our current system supports lap-top PC and PHS (Personal Handy phone System) mobile phone as communication devices for mobile users. PA converts terminal mobility information of PHS to user's location information. Also it is managed in the area agent (AA). AA manages the information about who are in that area. In cooperation with AA, PA will detect available communication devices at the user's location, and keep the data. Therefore, if you want to access a user, you can easily access the user through their PA. For example, shown in Figure 8.6, when a user is using a PC with Internet phone and the target person only has a PHS mobile phone, then PA recommends 'talk' method. This function is realised by setting up network resource and application automatically according to PA's user profile.

Also the data flow model using Pathwalker is easily applicable to the service routing functionality of PA. User profile contains the rule set (i.e. routing policy) which determines the data flow of event transactions and various users' statuses. Dispatcher in PA can determine the transaction according to user profile. For example, the PA receives schedule update events from calendar service, where the user's routing policy is 'if his/her PC is connected to a network, notify it by the dialog message of PC; otherwise, notify it by short message on mobile phone'. Personal events related to mail receiving, or schedule arrangement, are forwarded to the optimum communication device by PA. In the case of mobile phone, the contents may be converted to a short message.

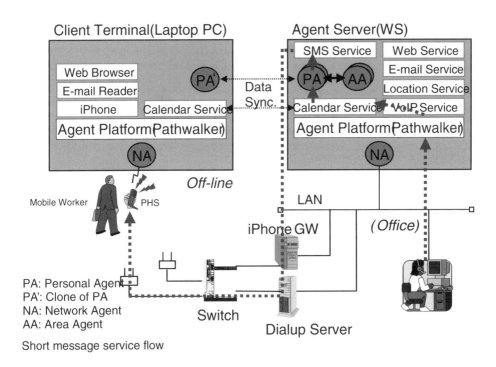

Figure 8.5 System configuration of VHE prototype

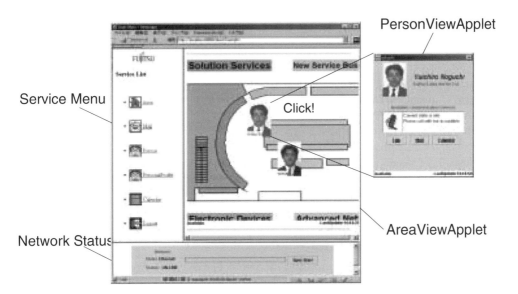

Figure 8.6 GUI views of map-based information service

Network agent (NA) resides in the client terminal; it can set up the network configuration automatically. For example, when a mobile user outside the office wants to access the intranet (office network) by dial-up sequence, NA detects PHS modem card as available medium and starts dialling to establish the network connection. When NA detects ethernet card as new available medium, it can set up TCP/IP to establish the network.

Clone of PA resides in client terminal as data cache; it stores up modified data while terminal is separated from the network. It automatically starts to synchronise with PA on the server side after the network connection has been established by NA. Therefore users don't need to worry about consistency of data between server and client terminal.

In this way, our prototype system provides a customised service using a user's routing policy and is a useful environment for mobile users. Soon this system will be realised as the product of the mobile middleware.

8.7   Conclusion

We proposed the multi-agent solution using personal agent (PA), network agent (NA), and radio agent (RA) as the components for Virtual Home Environment (VHE), then developed the prototype system. The PA that is used by routing policy in user profile can automatically select optimum communication method corresponding to the user's situation. NA can provide network plug and play and data synchronisation for supporting mobile communication environment. RA can solve the problem about instability of radio channel. It was proved that this system is available in a mobile communication environment and can provide a seamless network.

8.8   References

Fujino, N., Tokuyo, M. and Iida, I. (1998) Adaptive Picture Quality Control by WWW Proxy Server Cooperating with Radio Agent, (in Japanese), *IPSJ SIG Notes*, Vol. **98**, No. **53**, 98-MBL-5, pp.1-6, May.

Hartmann, C. Gorg and Farjami, P. (1998) Agent Technology for the UMTS VHE Concept, *Proc. of ACM/IEEE MobiCom'98, Oct. 1998.*

Iida, I., Nishigaya, T. and Murakami, K. (1995) DUET: An Agent-Based Personal Communications Network, *IEEE Communications Magazine*, pp. 44-49, Nov.

Imielinski, T. and Korth, H. (1996) *Mobile Computing*, Kluwer Academic Publishers, 1996.

Ushijima, S., Mohri, T., Iwao, T. and Takada, Y. (1998) Pathwalker: Message-Based Process-Oriented Programming Library for Java, *Proc. of INAP'99*, pp. 137-143, September.

8.9   Further Reading

Iwao, T., Okada, M., Takada, Y., Amamiya, M. (1999) Flexible Multi-Agent Collaboration Using Pattern Directed Message Collaboration of Field Reactor Model, *Proc. of PRIMA'99*, December.

Nishigaya, T. Design of Multi-Agent Programming Library for Java.
    http://www.fujitsu.co.jp/hypertext/free/kafka/paper/

9

Virtual Home Environments to Be Negotiated by a Multi-Agent System

S. Lloyd, A. L. G. Hayzelden, L. G. Cuthbert

9.1   Introduction

Convergence of once fragmented industries such as telecommunications, entertainment and its delivery, cable TV and computing continues to challenge the network operators and service providers as they evolve. Third-generation mobile, UMTS is likely to rapidly accelerate the convergence of fixed and mobile telecommunications. UMTS is a technology that builds on and extends the capability of today's mobile technologies by providing increased capacity, data capability and a far greater range of services over both fixed and mobile networks. As a result of greater bandwidth and enhanced network intelligence users will increasingly want advanced data and information services (Mobilennium, 1998). We believe services drive the evolution of a network. In converging networks voice communication may remain the dominant service in the foreseeable future, but other more advanced services will continue to increase in importance. Ultimately, a service goal of any network operator or service provider, thinking globally, is to provide universal, seamless services. Seamless services directly affect users: they are types of services that do not change their behaviour when users travel through network boundaries and different operating environments. Telecommunications will become a combination of fixed and mobile services to form a seamless end-to-end service for the user. One of the key service goals being introduced through the UMTS network is the Virtual Home Environment (VHE), which will attempt to make the 'user experience' of the services identical and seamless, where or however access is made.

In this chapter we propose a Multi-Agent System (MAS) to establish a service in a VHE using stationary, intelligent agents. One fundamental problem occurs when the local networks do not support the services that a user has subscribed to in a home network. A multi-agent system approach enables service providers, network operators and a VHE to be represented within the MAS. Negotiation between the agents can provide service functionality if local network or terminal equipment should need to be dynamically reconfigured or agreement is required to inter-operate with a new network operator or service provider. Another problem that inhibits the take-up of a VHE is the difficulty that users have in dealing with different access techniques for the same services when accessed from different networks. The MAS approach resolves any system issue conflicts and the access difficulties remain transparent to the user in the VHE. The VHE domain has many legacy systems that need to have their functionality extended. The MAS can provide another level of agent interaction for the legacy code to support the VHE. There are many technical and commercial developments that will be necessary for full implementation of the VHE. We believe we can exploit the benefits of agent technology to enhance the services offered by a VHE (Lloyd and Pearmain, 1999).

The contribution of this work is the application of a multi-agent system with stationary, intelligent agents to solve a difficult, real-world, telecommunications problem. The VHE is a complex, distributed application domain with distributed data, control and expertise.

The remainder of this chapter is structured as follows. We begin by introducing UMTS in section 9.2 and then describing the key components and expectations of a VHE in section 9.3. Section 9.4 explores the multi-agent system approach and the benefits it will offer. This is followed by section 9.5, which introduces Intelligent Networking (IN) principles into the VHE provision. The proposed agent architecture is detailed in section 9.6 and some conclusions are presented in section 9.7.

9.2   UMTS

Universal Mobile Telecommunications System (UMTS) follows on from first - and second-generation mobile systems. First-generation mobile brought analogue voice communication and ensured mobility within a specific network area. Such systems include analogue cellular systems, cordless systems and various national paging systems. Second-generation mobile communications, which include digital systems such as GSM, underwent further development and provided users with international roaming, flexibility and digital security (DTI, 1997). UMTS is being promoted internationally by ETSI (European Telecommunications Standards Institute), UMTS Forum and the Commission of the European Community. It has the support of the major fixed and mobile network operators and the manufacturing community. Third-generation mobile will initially use the radio spectrum around 2 GHz partly allocated globally and partly allocated regionally and will be the European implementation of a global family of mobile systems known collectively as IMT 2000 'International Mobile Telecommunications 2000.'

UMTS is a service-driven standard. The key differentiators between UMTS and second-generation systems are:

- combined and consistent access to services from all networks;
- personalisation of service to customers through adaptive terminals;
- open systems for the creation of new services and applications on demand;
- support for multimedia services with high-capacity voice, data and video support.

UMTS is conceived as a global system, comprising both terrestrial and satellite components. With new terminals a subscriber will be able to roam between private and picocellular/microcellular public networks, then into a wide-area macrocellular network and finally into a satellite mobile network (see Figure 9.1). This could be followed by a complete 'in call' change-over onto a fixed network during a call with identical service representation. UMTS will be the first system to offer mobile users roaming during an existing connection with hand-over between networks with different applications and different operators with no discontinuity of service. It will provide an open architecture that will facilitate the easy introduction of advances in technology and different applications.

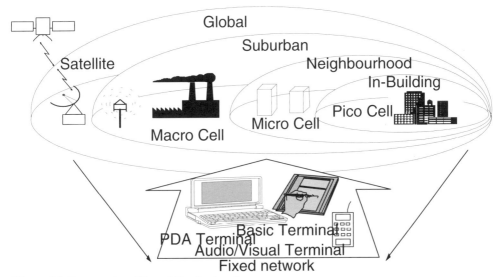

Figure 9.1 Integrated mobile and fixed networks

Fixed-network mobility is expected to have developed considerably by the time UMTS is introduced. Personal numbering and personal profiles will be available as customer services; these will lead to the idea that customers on fixed networks will no longer be identified primarily by a particular geographical location or by a physical network termination point but instead by a personal identifier. As fixed and mobile convergence continues, VHE will be the ubiquitous service environment for the user.

9.3   What is Virtual Home Environment?

VHE allows users to set up their own personalised service portfolios and use them in any other network. For a VHE it is envisaged the user will have a terminal that may be designed for mobile or fixed use. The user will be able to customise services and their presentation, and will be able to maintain these even when roaming to other networks on a global platform (Mobilennium, 1998; Lloyd and Pearmain, 1999;Torabi and Buhrke, 1998). UMTS will offer a consistent service through a Virtual Home Environment. The expectation will be for a range of services each selected in a uniform way with consistent procedures, performance and quality irrespective of the:

- means of access (mobile or fixed);
- application (cellular, cordless, paging, satellite, etc.);
- service provision (public or private);
- environment (home, street, transportation, office, shop, etc.);
- location (service area or global);
- terrain (urban to rural to mountain).

When roaming between networks (fixed and mobile) a personal user will experience a consistent set of services, thus 'feeling' as if they remain on their home network. The user will appear to have identical services offering familiarity and ease of use. Currently if a user is on a foreign network only the local-network operator functionality and 'look and feel' is experienced or a subset of the features is available. For example, operator services for national telephone number queries are available by dialling '192' in the UK; when roaming in other networks or countries a different number has to be used or the user has to choose the international queries for a UK number. Using the same number for the same service is only possibly with a VHE.

From the user's perspective the VHE scenario enables transparent access to subscription-based services when roaming. There will be one global access environment for the user that will be familiar, simple, and effective; to the user it will appear identical. For example, in her home environment Stephanie subscribes to a service that allows one-touch dialling to order a taxi. The information is either passed through a voice communication or by text-based responses. When Stephanie is roaming in Portugal she needs to order a local taxi. Stephanie has no knowledge of local taxi services or numbers. The VHE, knowing that Stephanie is roaming, starts to negotiate with local service providers and network operators for a local taxi service to support Stephanie's request. The network limitations, local services support and terminal interfaces are all controlled through the VHE, transparent to Stephanie.

From the terminal and network perspective, the VHE implementation scenario may require many technical developments and commercial agreements. To demonstrate the complexity, we have characterised the VHE into component areas:

- *Functionality* – this will be negotiated with the visited network so that it will provide 'home-like' service in a potentially alien environment.
- *Feature transparency* – this is the need to hide details of what kind of network provides the service at any one time to provide the same services, globally, in a consistent environment to the user.
- *Roaming* – a critical service or part of a service will be the ability to roam. Roaming occurs when a user travels outside his or her home serving area and requests service.
- *VHE standard interfaces (signalling and protocols)* – open standards complemented by proprietary service differentiation will be needed to enhance but not jeopardise competitive advantages for service providers and network operators (e.g. Internet).
- *Distributed intelligence* – there are levels of access to the intelligent functionality embedded within Public Telecommunication Operators (PTOs) which will need to be fully distributed and not located only centrally or peripherally.

- *Network access* – this plays a pivotal role in providing VHE services. The capabilities of network access interfaces largely determine what kind of services can be offered to users. As an example, if the network is capable of producing information for a customer, but an access interface is not capable of using this information or delivering it to a customer, no service relying on this information can be supported.
- *Optimal routing* – this functionality can be used for both voice or data and signalling messages for the VHE. A current example could be: if Johanna, a customer with home network in Holland, calls Robert, an English customer also roaming in Holland, the call is always routed via Robert's home network in England. For the users, both Johanna and Robert end up paying for international call legs even though the called parties are actually in the same network (tromboning). Network capacity and resources were used that could have been deployed elsewhere.
- *System issue resolution* – system issues arise when there are differences that exist between similar systems deployed by different networks. For example, databases in two networks contain the same type of information, e.g. service profile of users. The database in one network generates a specific command to another network element (e.g. a switch) when it is queried; if the database in the other network does not do that the user will see some difference in the service. The VHE would need to resolve any conflict before delivery of service.
- *Personal profiles* – a profile should define the services required and these should be maintained even when roaming to other networks.
- *Service efficiency* – this will be needed in order to provide a service efficiently, e.g. to cross-use databases located in adjacent networks.
- *Maintenance of transient data* – data will need to be kept in synchronisation with data held by the service provider of the VHE.
- *Network control* – this will offer even distribution and control over management, bandwidth allocation and use of resources, particularly under failure conditions.
- *Common service processing* – this may include registration, interpretation of addressing, interpretation of service logic and interpretation of the user interface data into session control information or service provider dialogues.
- *Common call processing* – call origination and registration, call control and call termination.

These components show that the VHE vision encompasses many areas, which can also be applied to other services. Improvements are currently being addressed and developed through the use of Intelligent Networking (IN) (Magedanz, 1996), Customised Applications for Mobile network Enhanced Logic (CAMEL) (Becher, Klas, and Leitgeb, 1997; Ambrosch, Maher, and Sasscer, 1989) and distributed intelligent agents for the management of telecommunications networks (Busuoic, 1999; Albayrak, 1998a). An MAS used to establish the VHE would complement these and existing mechanisms where possible. VHE needs to be part of the network infrastructure and will need to have functionality within network nodes. UMTS is promoting VHE but it is our view that the VHE will ultimately operate within the network infrastructure, inter-working over many network and transport protocols.

9.4   Agent Rationale

The previous section produced a comprehensive view of the requirements of the VHE and therefore also provided example areas that agent technology can enhance. For establishing the VHE we decided a distributed software agent approach has particular value. For the general advantages that agent technology has brought to complex software systems development we refer to (Jennings et al, 1996; Jennings, 1996). Extensive definitions and overviews of agents can be found in Jennings, Sycara and Wooldridge (1998), Nwana (1996) and Torabi and Buhrke (1998) and their application in the telecommunications field is well documented in Albayrak (1998a), Hayzelden and Bigham (1999), Hagen, Breugst, and Magedanz (1998). There are various reasons for choosing the use of agent technology, but a particularly strong argument is the use of agents for conceptualisation of system design. At a high level the multi-agent systems approach is intuitively simple: developers can draw on their experience in solving real-world problems in co-operation with others. This gives software agent developers an advantage when constructing multi-agent systems, as there is some reliance on a natural methodology for isolating the different interaction modules that make up the complete system.

Table 9.1   VHE attributes matched to agent roles/technology

VHE Attribute	Agent Roles/Technology
Optimisation of routing (between home and visited networks)	Global and local decision making
Provide service functionality (from the earlier example when local switch does not support the '192' number translation function)	Broker and bid for services and resolve conflict (pay for different levels of QoS), explore interdependencies
Robustness of the service	Distributed framework to tolerate failures
Performance targets for users, service providers and network operators	Goal oriented to both reactive and deliberative tasks
Control of resources (reconfigure transmission links)	Invoking changes without the need for human interaction
Feature transparency (user sees no difference)	Negotiation of network resources – select optimum communication device
Complex application domain	Several components handled by separate agents, natural model of computation
Common service processing across VHE	Provide ontology and conflict resolution to discrepancies by co-operation
Distributed intelligence across many network elements	Distributed rational decision-making using strategies
Service efficiency	Prediction through user trends and dynamically adapt
Continual improvement of VHE services to the user	Learning and adapting service logic
Inter-working with legacy software	'Wrapping' with agents

The agent approach also provides an appropriate analogy for the decomposition of problems and concurrent task delegation by supporting the abstraction of some of the complexities that are present in the telecommunications domain. Table 9.1 shows how agent roles and technology can benefit and improve the key VHE attributes and therefore how a VHE implementation is an ideal application for a multi-agent system.

Currently there are projects investigating mobile agent-based services in fixed and mobile telecommunication environments (Breugst et al. 1999; Pham and Karmouch, 1998; Gervais and Diagne, 1998). However, mobile agents require a highly secure execution environment, which could limit their performance and functionality in telecommunications applications (Harrison, Chess and Kershenbaum, 1995). An overhead for the messages with intelligent agents does exist but in a telecommunications environment this may not outweigh the security issues and overheads raised by using mobile agents. Therefore, it is proposed to develop the solution using stationary intelligent agents.

9.5   Intelligent Network as a Trigger to the VHE

GSM has successfully automated inter-operator roaming with the provision of basic standardised service sets (Clapton, 1998). Experience of GSM has shown that it is vital to be able to introduce new services without excessive delay or disruption and the GSM system specification has enabled operators to rapidly deploy feature-rich networks; however differentiation through original services has been reduced. The GSM community has addressed this problem via the development of the GSM CAMEL (Customised Applications for Mobile Network Enhanced Logic) feature (Becher, Klas, and Leitgeb, 1997). CAMEL is the integration of Intelligent Network (IN) techniques into the GSM network that will enable an operator to provide specific services when a subscriber is roaming. CAMEL uses an evolved IN capability (Ambrosch, Maher, and Sasscer, 1989).

The IN concept has been a major driver in the evolution of telecommunications networks during the last decade with most major network operators deploying IN-compliant platforms. It is proposed that the IN will act as an intermediate layer between converging telecommunications networks to provide an interface to a multi-agent system for the VHE.

IN functionality is being developed by both fixed and mobile network providers to separate the mobility and service layer from the switching layer. One of the prime technical aims of UMTS is to separate out the mobility and service support layer from the transport and switching layer (Cullen and Lobley, 1996). The IN architecture provides several important advantages, like the opportunity to create or modify network capabilities without any changes at the switches. This provides speed of deployment and flexibility to the network operators. However,because of the rising number of users and the increasing number of IN services, improvements are required. The IN architectures are effectively centralised, complex and difficult to handle (Albayrak, 1998b). The functionality in the areas of service creation and management is insufficient and real interoperability of services is difficult to achieve. IN development using agents has already been undertaken (Magedanz, 1996).

9.6   Proposed Agent Architecture

The agents represent the key roles, at a high level, within the telecommunications environment to provide the VHE to the user. The high-level multi-agent system will include a customer terminal agent which will be housed in each terminal (see Figure 9.2). The terminal will be used for fixed or mobile service. If the terminal is switched off or loses a connection during mobility then the control will automatically be given to a customer network agent. The customer network agents will then negotiate on behalf of the customer terminal agent until the mobile is back on-line and gains control. When a VHE is requested by either of the customer agents a VHE controller agent will negotiate to enable the particular service, made up of features, to be delivered (always to the customers 'look and feel'). This can be negotiated from the home or preferably the visited networks. An agent represents each service and each feature. Negotiation between the agents can provide service functionality if local network or terminal equipment should need to be dynamically reconfigured or agreement is required to inter-operate with a new network operator or service provider. The agents resolve any system issue conflicts, and the access difficulties that users have in dealing with different access techniques for the same services when accessed from different networks, remain transparent.

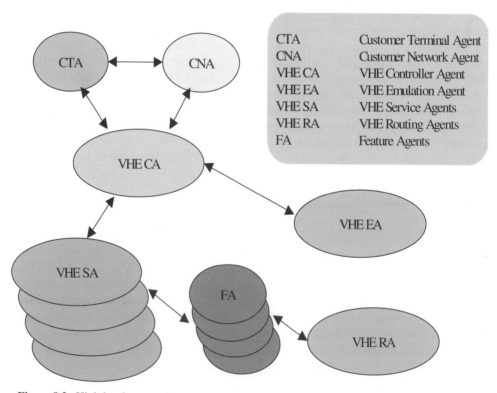

CTA	Customer Terminal Agent
CNA	Customer Network Agent
VHE CA	VHE Controller Agent
VHE EA	VHE Emulation Agent
VHE SA	VHE Service Agents
VHE RA	VHE Routing Agents
FA	Feature Agents

Figure 9.2   High-level agent roles

A feature agent may incorporate a routing function. Routing will need to be optimised; especially if a call needs to be routed back to a home network. If required within a feature agent then a VHE routing agent will be used to find the most effective path. Due to the complexity of routing, separate agents manage this entity. An emulation agent will be working for each VHE controller agent because if negotiation fails to provide service as requested the emulation agent is responsible for providing a backup for the user with the consistent 'look and feel'. Service providers and network operators are represented within the architecture as part of the VHE service agents. Service providers will negotiate deals with network operators for transmission paths, for example. The service layer interfaces to, and makes demands on, the transport layers. For example, a service contract (described by a Service Level Agreement) might stipulate a specific Quality of Service, cost and penalties and even some routing request (e.g. not via country X for security reasons or to avoid encryption regulations). The service layer may invoke routing 'rules' if necessary but would not be involved with the physical packet routing itself, which is a job for the transport layers and the function of the network operator (or reseller).

The VHE controller agent will continue to negotiate with the local service agents to provide the features for the next request for the VHE, even though emulation may have been needed in a previous instance. The immediate goal of the VHE controller agent will be to manage a VHE service agent to provide a consistent service. VHE controller agents will be located at the service provider. Service and feature agents could be service providers or network operators.

The VHE is being developed using a Foundation for Intelligent Physical Agents (FIPA) Open Source (OS) agent platform (FIPA-OS, 1999), which is an environment where agents are deployed, managed and executed. This platform was chosen for a number of reasons. By conforming to the FIPA standards we are creating an open environment, where other applications built using FIPA-OS Agents, should be able to inter-operate. We are using the rich set of resources the platform provides to facilitate the operations of groups of agents and, finally, in addition to the mandatory components of the FIPA Reference Model, the FIPA-OS distribution includes an Agent Shell, and an empty template for an agent. Multiple agents can be produced from this template, which can then communicate with each other using the FIPA-OS facilities. This also speeds the development process since infrastructure for facilities such as agent messaging are provided.

9.7   Conclusion

This chapter has focussed on using a multi-agent system to design a VHE in an environment for converging networks. We have analysed possible components for a VHE deployed in converging networks and shown that the application domain is suitable for an agent approach. The main technical problem with a Virtual Home Environment is to provide the same services, globally, in a consistent environment. VHE could be achieved using a set of protocols but they remain fixed and may not have the functionality, framework or flexibility that a multi-agent system offers. An idealist model suggests a vision of a fully interactive and seamless service environment in which access, via public networks, to all conceivable electronic services is available to all users at ever-declining prices. The telecommunications establishment is often keen to sustain the myth of ubiquitous, flexible, open public telecommunication; however, some of the issues raised in this chapter will need further breakdown of legacy telecommunications operator barriers to fulfil a vision of VHE services using a multi-agent system.

9.8   References

Albayrak S. (ed.) (1998) *Intelligent Agents for Telecommunications Applications*, IOS Press, ISBN 90 5199 295 5.

Albayrak S. (1998b) Introduction to Agent Oriented Technology for Telecommunications, Chapter 1 in *Intelligent Agents for Telecommunications Applications*, Edited by Albayrak S., IOS Press, ISBN 90 5199 295 5

Ambrosch W. Maher A. and Sasscer B. (1989) (eds.) *The Intelligent Network (A Joint Study by Bell Atlantic)*, IBM and Seimens, Springer Verlag, ISBN 0 387 50897

Becher R., Klas G. and Leitgeb M. (1997) CAMEL: The impact of personal communications on Intelligent Networks, *Proceedings of International Switching Symposium, Toronto, Canada*, pp. 225-233.

Breugst S. et al. (1999) Grasshopper – An Agent Platform for Mobile Agent-based Services in Fixed and Mobile Telecommunication Environments, Chapter 14 in *Software Agents for Future Communication Systems*, Edited by Hayzelden A. and Bigham J., Springer Verlag, ISBN 3 540 655786

Busuoic M. (1999) Distributed Intelligent Agents – A Solution for the Management of Complex Telecommunications Services, Chapter 4 in *Software Agents for Future Communication Systems*, Edited by Hayzelden A. and Bigham J., Springer Verlag. ISBN 3 540 655786

Clapton A. (1998) UMTS – the mobile part of broadband communications for the next century, *BT Technology Journal*. **16(2)**. April, pp. 120-131

Cullen J. Lobley N. (1996) The Universal Mobile Telecommunications System – a mobile network for the 21st century, *BT Technology Journal*. **14(3)**. July, pp. 123-131

DTI (1997) Developing Third Generation Mobile and Personal Communications into the 21st Century, 31 May 1997, http://www.gtnet.gov.uk/radiocom

FIPA-OS (1999) http://nortelnetworks.com/fipa-os

Gervais M. and Diagne A. (1998) Enhancing Telecommunications Service Engineering with Mobile Agent Technology and Formal Methods, *IEEE Communications Magazine*, July pp. 38-43.

Hagen L. Breugst M. and Magedanz T., (1998) Impacts of Mobile Agent Technology on Mobile Communication System Evolution, *IEEE Personal Communications*, August, **5(4)** p. 55-69.

Harrison C., Chess D. and Kershenbaum A. (1995) Mobile Agents: Are they a good idea? Research Report, IBM Research Division, NY, http://www.research.Ibm.com/massive/mobag.ps

Hayzelden A. and Bigham J. (1999) Agent Technology in Communications Systems: An Overview, *Knowledge Engineering Review*, **14(4)**.

Jennings N.R. et al. (1996), Agent-based business process management, *International Journal of Cooperative Information Systems*, **5(2-3)**:105-130.

Jennings N.R. (1996) Using Archon to develop real-world DAI applications for electricity transportation management and particle acceleration control. *IEEE Expert*, December, **11(6)**: pp. 60-88.

Jennings N.R., Sycara K. and Wooldridge M. (1998) A Roadmap of Agent Research and Development. *Autonomous Agents and Multi-Agent Systems*, Edited by Jennings N. Sycara K. and Georgeff M., **1(1)**, pp. 7-38, Kluwer Academic Publishers

Lloyd, S. and Pearmain, A. (1999) Multi Agent System for Establishing 'Virtual Home Environments' in the Convergence of Fixed and Mobile Telecommunications Networks'. *AAAI Workshop 'Artificial Intelligence for Distributed Information Networking*, AAAI Press, July 1999.

Magedanz T. (1996) Mobile Agent-based Service Provision in Intelligent Networks, IATA Workshop at ECAI.

Mobilennium (1998) The UMTS Forum Newsletter, No. **5** Nov. Available at http://www.umts-forum.org/mobilennium.html

Nwana H. (1996) Software Agents: an Overview, *The Knowledge Engineering Review*, **11(3)**, pp. 205-244

Pham A. and Karmouch A. (1998) Mobile Software Agents: An Overview, *IEEE Communications Magazine*, July p26-37

Torabi M. and Buhrke R. (1998) Third Generation Mobile Telecommunications and Virtual Home Environment: A Prioritization Analysis, *Bell Labs Technical Journal*, July-September, pp. 50-62.

10

Self-Adaptation for Performance Optimisation in an Agent-Based Information System

C. Gerber

10.1   Introduction

Over the last years, individual mobility has increased significantly: a growing number of people travel over more and more expansive distances, leading to an enormous demand for means of transportation. Partially, this demand is met by the construction of new roads and the expansion of railway and airway systems. However, such extensions cannot catch up with the increase in demand. Hence, new approaches must be introduced to enable more efficient use of already existing transportation means.

A second phenomenon of the 1990s is the tremendous increase in electronically available information, mainly because of the explosion in performance of computer hardware, the booming use of wireless communication facilities and the broad acceptance of the Internet. These new media allow for the use of integrated information systems in telematics applications of dimensions not conceivable some years ago.

The MoTiV-PTA[1] project has been initiated by the German Ministry for Education, Science, Research and Technology (BMBF) in order to develop a unified distributed telematics system for planning and supporting individual inter-modal travel. Industrial partners are BMW, Bosch, Daimler-Chrysler, IBM, Opel, Siemens, VDO, and Volkswagen. The project is being developed for personal and individual support of travellers. Such an assistance includes route planning (for multiple means of transportation), accommodation and dining booking, reservation of tickets for cultural events, etc.

During a trip the MoTiV-PTA system will not only assist a user with car navigation, it will also accompany them while using basically any transportation system. In particular, the user of such a system will be able to get on-line information on different means of

---

[1] German acronym for *Mobility and Transport in Inter-modal Traffic - Personal Travel Assistance*

transportation, accommodation, etc., obtain a detailed proposal of a complete itinerary with the best expected user satisfaction (including time schedule and route plan, hotel suggestion, etc.), and conveniently book the chosen means of transportation and accommodation.

10.2   The MoTiV-PTA System

Three types of user interface devices can be used: a notebook-like tool, called *Mini-Personal Travel Manager* (PTM) or *Personal Intelligent Communicator* (PIC) where the user accesses the system via Windows CE-based devices running a dedicated graphical user interface, a system to be integrated in the user's car dashboard and a piece of software which is run on the user's PC. The system running on a PC is intended to have all necessary capabilities for off-line performing of the tasks required for a travel manager. The capacities of PICs and dashboard tools, on the other hand, are much too small to perform complex planning and negotiation processes. They are designed for mobile on-line usage: these tools communicate with more powerful mirror modules over GSM or ISDN.

The integrated information-providing services are distributed on different servers, accessible via Internet. The key idea here is to provide an interface to already existing information sites in order to allow for automated information retrieval. The MoTiV-PTA specification contains inter-modal route planning combining different transportation modalities such as railroad, automotive and airways, a combination of automobile route planning with car park scheduling, traffic monitoring, hotel information and reservation, and tourism information.

10.2.1   Architecture of the MoTiV-PTA System

The MoTiV-PTA system is implemented in Java for platform-independence reasons. The user accesses the system via a *user device agent* (UDA) running on their local device. The UDA communicates via ISDN or GSM with a MoTiV-PTA server. Here a *dispatch agent* distributes the incoming messages to the corresponding *user agent* (UA) of the UDA. The UA manages the user preferences and sends requests taking the preferences into account to a *broker agent* to perform a complete planning of a journey.

The broker agent decomposes the request into requests for individual service agents, such as an inter-modal route planner (IMRP) for calculating the route or a car park manager for selecting a car park. For achieving up-to-date information, this agent contacts information providers, for instance agentified railway or airline information systems. The broker agent combines the results and passes the alternative itineraries to the user agent. This agent sends the results to the user device agent which presents them to the user.

The user chooses between the prepared alternatives, the chosen itinerary can then be monitored for unforeseen events, such as traffic jams on the scheduled route or the chosen parking lot becoming full. If the current itinerary is no longer feasible, re-planning is automatically initiated, and the user is informed of the consequences. Figure 10.1 depicts the MoTiV-PTA agent society.

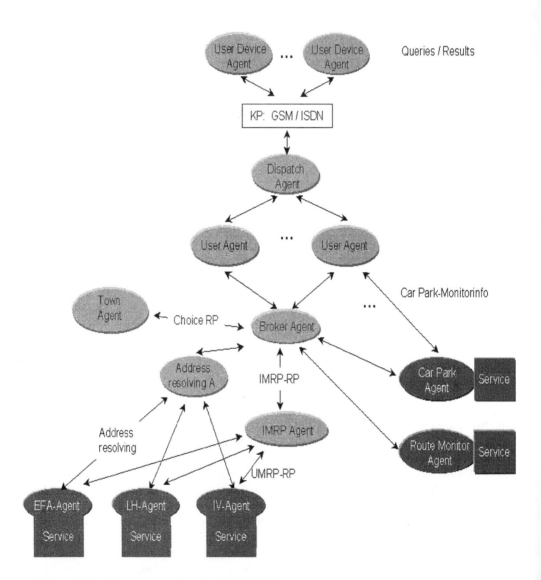

Figure 10.1 The agent society in MoTiV-PTA

Before a system such as MoTiV-PTA can be introduced into the market place, a series of test runs and simulations have to be performed. The agent-based simulation framework *SIF* (Funk et al. 1998) is used to achieve a successful application in two ways. First, it is used to build a simulation engine and to determine the optimal system configuration for a given environment profile. In section 10.3 we show how the simulation engine is set up. Second, SIF is used in on-line simulations (and later, in realistic runs) to adapt the system to dynamically changing environments and to achieve scalability. The overall optimisation task is thus to minimise the communication effort and run times of the various services such as information retrieval, bookings, etc. in the MoTiV-PTA system. Since the types of services may vary and the setting of the network may also diverge greatly, we focus on optimising the average duration of services during the daily run of the system. In section 10.4, we present the optimisation procedure in more detail.

10.3   A Simulation Engine for MoTiV-PTA

In the following we describe all extensions to the existing MoTiV-PTA prototype to a sufficiently realistic simulation testbed. The key idea for the simulation is to introduce agents of the SIF agent architecture for an explicit representation of hardware servers. First, we sketch the basic principle of SIF, then we give an overview of the implemented simulation engine and finally describe its user interface.

10.3.1   The social interaction framework (SIF)

The MAS simulation environment SIF has been developed in Java for the study of social interaction between artificial agents of arbitrary architectures. SIF's underlying basic mechanism is a so-called *Effector Medium Sensor* (EMS) architecture based on Russell and Norwig's definition of an agent in Russell and Norvig (1995).

An agent is anything that can be viewed as perceiving its environment through sensors and acting upon that environment through effectors.

The central component of the architecture is the *medium. Effectors* emit actions to the medium, which in turn sends the effect of that action as *percepts* to *sensors* (Figure 10.2).

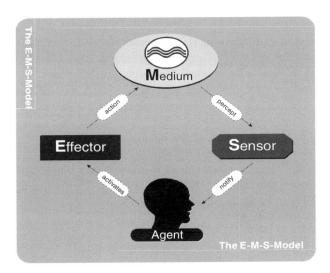

Figure 10.2 Basic principle of SIF

Agents are entities equipped with effectors and sensors in order to emit actions to and receive percepts from the *world server*, an instantiation of the medium, representing the environment. Hence, agents perceive other agents' actions indirectly in the world server: 'filters' the effects of actions by computing the local perspectives of perceiving agents. The world server sends these perspectives as percepts to the sensors of perceiving agents (Figure 10.2). Details of the SIF framework can be found in Schillo et al. (1999); the SIF development kit can be downloaded for free under the GNU licence agreement from *http://www.dfki.de/~sif.*

10.3.2   Architecture of the simulation engine

For a realistic simulation we represent explicitly the hardware servers and communication duration. Additionally, we simulate the services provided by the MoTiV-PTA prototype that runs on these servers, because the overall performance of a MoTiV-PTA system does not only depend on the efficiency of the agent architecture but is also heavily influenced by the following factors:

- The user behaviour may vary over location and time: certain types of requests may be initialised more often from one part of the domain than others. Additionally, there may be a work load distribution on a temporal basis.
- Servers might be overloaded or might break down; usually they vary in terms of equipment and are geographically dispersed.
- Depending on the type of the communication medium, e.g. Internet, GSM-bearer, or ISDN, the medium throughput is different.
- The duration of services may also differ. This depends on the server used, but also on the work load of the service providers induced from other non-PTA-based sources (e.g. a railway information provider may also be accessed by a vast number of travel agencies).

The above factors lead to different service run times, dependent on the work load and the configuration of the servers the MoTiV-PTA agents are logged on. In order to integrate these factors into a simulation engine, we apply modifications to the MoTiV-PTA prototype. However, we want to interfere with the original MoTiV-PTA system as little as possible in order to avoid side effects between the simulation engine and the system to be evaluated.

Figure 10.3 shows a screen shot of the integrated simulation system: on the left, the original MoTiV-PTA agents are shown while the right-hand side displays the console simulation engine (indicating the locations of the servers on a rough map of Germany) and an information window that indicates the state of a server. The following extensions of the original MoTiV-PTA architecture are made.

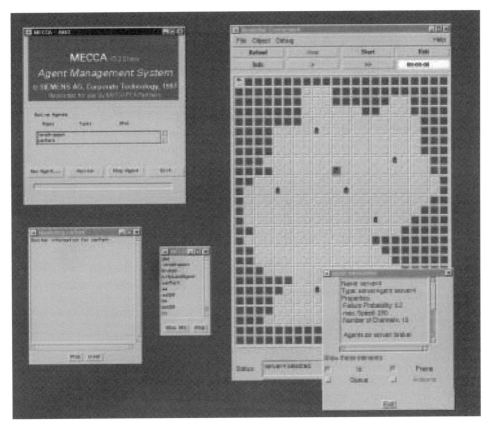

Figure 10.3   Screen shot of the system

10.3.2.1   Explicit representation of servers

In a first step we extend the original MoTiV-PTA architecture by aggregating each MoTiV-PTA agent with an SIF agent representing a hardware server that implements the communication platform of a MoTiV-PTA agent (see Figure 10.4). We do not use simple objects, but autonomous agents for representing servers in the simulation in order to model more accurately server failure behaviour: an agent can autonomously determine and change its state according to the simulation parameters. The use of MoTiV-PTA agents would spoil the accuracy of the simulation since such agents have to register at the system to be simulated. We use the SIF framework for this purpose since SIF is especially designed for simulation purposes and also written in Java allowing an easy integration.

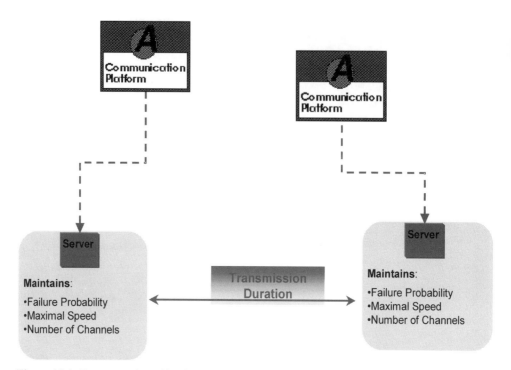

Figure 10.4 Representation of hardware servers

10.3.2.2 Integrated long-term simulation

Since there are many factors that influence run times in MoTiV-PTA, we provide an integrated long-term simulation environment, where user requests are posed according to a frequency distribution in terms of geographical distribution of the user, temporal distribution of the number of requests, and a distribution of the type of requests the user can submit. The process duration for these requests can be measured and used for statistical evaluations. SIF offers constructs for simulating these distributions: the SIF world server can easily be extended to control the timely emission of appropriate requests.

10.3.2.3 Simulation of delay of communication among agents

Each SIF server agent is assigned to a certain simulated position in the network. Whenever two MoTiV-PTA agents communicate, a logical distance between their two servers in the simulation is determined which need not correspond exactly to the geographical distance. Due to the structure of the Internet, transmission between very close sites is not necessarily faster than between distant sites. Communication duration has to be simulated depending on the logical distance of the server. We employ the SIF world server to control the duration of the communication acts.

10.3.2.4   Migration

We extend the scenario by enabling user agents to *migrate*. In a traditional approach, communication between geographically distributed agents is performed over the network. This can be rather time-consuming in the case when the network is heavily loaded and the communication process consists of some complex negotiation procedures. In the *migration approach,* an agent is transmitted over the net in order to communicate with its partner on the local server. A user agent, for example, might travel from home to a ticketing place to obtain theatre tickets. Later the agent might travel back home to describe to its user the tickets it has obtained.

This approach is fast if the receiving server has an accurate model of the travelling agent. In this case, only data describing the agent's state (i.e. the instance variables) have to be transmitted; a copy of the agent can be generated locally. In this simulation migration can be realised even more simply: an agent has to modify only its server agent link.

It may make sense to let only those agents migrate that are expected to do a lot of long-distance communication. Such a setting is given if, e.g., the user (and his or her UDA) have moved temporarily far from the location of the UA. For the communication between UDA and UA, distance is irrelevant, since it is done over GSM or ISDN; however, since many of the tasks the system performs are used for information retrieval (i.e. communication with service agents) of the new logical neighbourhood of the user, it is reasonable to allow the UAs to migrate to a server closer to the service agents and to integrate this feature into the simulation.

10.3.2.5   Registration of UDAs

The logging processes of UDAs to servers are also simulated: whenever the user emits a new request, the corresponding UDA which is located on the PIC or any other local system has to connect to the server of the user's UA which may be its home server or some other server if the user agent has migrated previously. If for some reason logging to that server is not possible, the UDA must log on to another server in the simulation; the choice is made on the distances and on the expected current speeds of the new candidate servers.

10.3.2.6   Simulation of services

All services in the simulation scenario are replaced by dummy procedures that do not actually provide the real service, but do determine a simulated service duration time on which the answer is delayed.

10.3.3   PTA-specific simulation constructs in detail

The following details of the simulation have been established in cooperation with experts from Siemens. The resulting values for service and transmission duration do in principle represent real-world circumstances. The achieved accuracy suffices for the realised prototype which has been developed to demonstrate the applicability of the approach.

A SIF server agent maintains the following instance variables:

- *position* = [*xPos;yPos*], *xPos, yPos* ∈ [0,20], the location of a server,
- *failureProbability* ∈ [0,1], the reliability of a server,

- *state* ∈ {down, up}, indicating whether the server is down at the moment (depending on the failure probability),
- *PTAAgentList*, a list of all MoTiV-PTA agents running on that server at a time,
- *numberOfChannels* ∈ [0;50], determining how many channels can be opened at most for connecting user device agents,
- *UDAList*, a list of all user device agents that are currently connected (up to the maximum number of channels),
- *maxSpeed* ∈ [0;400], the maximum speed of the machine in some abstract measure,
- *currentSpeed* ∈ [0;400], the current processing speed depending on the number of agents running on that server and maximal speed of the server:

$$\text{currentSpeed} = \text{maxSpeed} \times \min(1, \frac{5}{\text{numberOfChannels}})$$

The duration (in milliseconds) of a transmission from server s_1 to server s_2 is computed according the formula

$$\sqrt{(\Delta x\text{Pos})^2 + (\Delta y\text{Pos})^2} \times \frac{1000}{6} \times \left(2 - \frac{\text{currentSpeed}(s_1)}{400}\right)\left(2 - \frac{\text{currentSpeed}(s_2)}{400}\right)$$

In the case of communication between UDA and UA, a constant time of 2000 msec is used for GSM communication, which includes the establishment of the connection and the transmission of the message. In the simulation, the SIF world server controls the simulation time; only if the computed transmission time has passed is the communication act sent to the receiver's communicator.

We enable those UAs to *migrate* whose current distance to the corresponding service agent is greater than a certain minimum distance. A user agent can only migrate if the server in question is currently not down (which is determined by the corresponding SIF server agent). If the server is down, another server is selected for the attempt to migrate. The one server is selected that maximises a tradeoff between speed and distance to the broken server:

$$\frac{\text{currentSpeed}^2}{(\Delta x\text{Pos})^2 + (\Delta y\text{Pos})^2}$$

In the simulation, a MoTiV-PTA agent migrates from one server to another by simply exchanging the aggregated server agent and by modifying the *PTAAgentList* vectors of both server agents. The duration of a migration attempt is computed according to the following formula:

$$\sqrt{(\Delta x\text{Pos})^2 + (\Delta y\text{Pos})^2} \times 500 + \frac{10000}{\text{currentSpeed (target)}} \, [\text{msec}]$$

Logging of an UDA on a server is simulated similarly by changing the aggregated server agent from PIC to the selected server and by adjusting UDALists accordingly, if that server is active and has unused communication channels at its disposal. Otherwise, another server has to be found, just like in the migration case. The duration of that procedure is computed similarly to migration duration:

$$\sqrt{(\Delta xPos)^2 + (\Delta yPos)^2} * 2000 + \frac{10000}{currentSpeed}[\text{msec}]$$

UDAs are modified to directly aggregate to the SIF system: each UDA holds a reference to the world server which forces the UDA to emit a request from time to time. Furthermore, the UDA has to report on the reception of an answer to that request.

The answer of a *dummy service* is delayed according to an abstract *serviceProviderSpeed* $\in$ [0;500] and the current *numberOfRequests* $\in$ [1;∞[which indicates the number of requests a service source is currently processing. These requests need not necessarily all be imposed by the MoTiV-PTA system, but can also be posed through other media, such as call centres.

The world server can modify the number of requests during the simulation run. The world server simulates the process delay for a service according to the formula

$$\frac{500}{serviceProviderSpeed} \times \log_{10}(numberOfRequests) \times 1000[\text{msec}]$$

10.3.4   *Functionality of the extended scenario script parser*

The MoTiV-PTA system has been enriched with a script parser that comes with SIF in order to allow the specification of scenario parameters that do not vary during a run of the simulation. This way we need not always modify and re-compile the main source code for scenario changes. We now briefly describe the elements of the simple script language.

```
Country Germany
Migration 22
ServerAgent server1 14 24 0.1 100 25
ServerAgent server2 11 10 0.2 80 15
ServerAgent server3 11 22 0.15 150 20
ServerAgent server4 17 9 0.05 200 10
Target server1

DF df server1
AMSAgent ams df server1
IMRPAgentWrapper imrpWrapper ams df server3 300 200
BrokerAgent broker ams df server9
CarParkAgent carPark ams df server1 250 150
UserDeviceAgent johnUDA ams df
UserAgent john johnUDA ams df server1
UserDeviceAgent maryUDA ams df
UserAgent mary maryUDA ams df server1
```

Figure 10.5 A sample scenario script

10.3.4.1   Relevant entities for the SIF simulation environment

- *COUNTRY* <*country*> A script line beginning with the key word COUNTRY determines the general type of the scenario. The user can specify a country; an abstracted map of that country will be then be loaded and displayed in the main simulation window (see Figure 10.3 which displays a map of Germany).
- *SERVERAGENT* <*serverName*>, <*xPos*>, <*yPos*>, <*failureProb*>, <*maxSpeed*>, <*numberOfChannels*> A script entry beginning with the keyword SERVERAGENT leads to the creation of a SIF server agent named <*serverName*>, located in the map at position [<xPos>;<yPos>] and equipped with the specified failure probability, number of channels and maximum speed.
- *MIGRATION* <*number*> This parameter defines the initial value of *minimum migration distance*. Only user agents that are located farther to the target than this value are allowed to migrate.
- *TARGET* <*serverName*> For the sake of visibility and simplicity we focus on the simulation of requests stemming from UDAs located on *different* servers which ask for information concerning only *one* location in the web. This can be specified by naming a *target server*. Requests to other locations could be simulated as well; however, results are rather independent from the simulation of requests aimed at other locations. Hence, we focus on only one target server.

10.3.4.2   Original MoTiV-PTA entities

- *DF* <*DFName*>, <*serverName*> The user can specify a *directory facilitator DF* (a yellow pages agent) which is simulated to run on a server named <*serverName*>.
- *AMSAGENT* <*AMSName*>, <*DFName*>, <*serverName*> Similarly, an *agent management system AMS* (which allows for convenient agent creation, deletion and monitoring) is created which runs on the specified server and registers at the named DF.
- *USERDEVICEAGENT* <*UDAName*>, <*AMSName*>, <*DFName*> A user device agent is created that is registered at <*DFName*> and <*AMSName*>. This user agent is initially not associated with a specific server since it is simulated to run on the user's individual device (e.g. a PIC).
- *USERAGENT* <*UAName*>, <*AMSName*>, <*DFName*>, <*UDAName*>, <*serverName*> A user agent is created which corresponds to <*UDAName*> and registers at the specified DF and AMS and runs on the given server.
- *BROKERAGENT* <*BAName*>, <*AMSName*>, <*DFName*>, <*serverName*> A broker agent is created which registers at the specified DF and AMS and runs on the given server. In analogy, agents of the type *ResolveAddressAgent* can be specified.
- *IMRPAGENTWRAPPER* <*IAName*>, <*AMSName*>, <*DFName*>, <*serverName*>, <*avgService-ProvSpeed*>, <*avgNumberOfRequests*> A dummy agent for inter-modal route planning is created. The last two parameters express the speed and work load of the service provider. Similarly, agent wrappers of the type *CarParkAgent, CarAgent, AirwaysAgent and RailroadAgent* can be specified that agentify different types of information services for uni-modal travel.

10.3.5   *User interface of the simulation engine*

Whereas scripts are used to load scenario parameters that do not change during the run of a simulation, SIF offers functionality, which can be modified to control an experiment during the actual run of the system. Additionally, a new graphical user interface, the simulation manager, helps to modify parameters during the actual run of the system. This module is aggregated to the world server.

10.3.5.1   Control features of SIF

- *Buttons*: The SIF console provides a number of buttons for user interaction. All buttons are active if their functionality is currently available. Pressing the start button activates the world server and the agents while pushing the stop button halts the computation. Using the reload button deletes all agents and resets the world representation of the simulation engine. Furthermore, the previously selected script is reloaded. Pressing the information button opens an object info window (see below) which contains information on a selected agent.
- *Menus*: The file menu offers general functions such as to load a new script or save the current state of the simulation as a script. By clicking on edit, scripts can also be modified. The object menu offers functionalities to kill previously specified agents and display information of the internal state of an agent.
- *Object Info Window*: For every server agent, this window displays static information like its name, position, failure probability, number of channels, maximum speed, but also dynamic information such as current speed, currently logged MoTiV-PTA agents and used communication channels. (Figure 10.3 shows such a window at the bottom of the right side.)

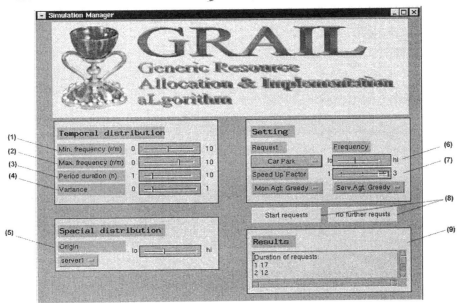

Figure 10.6 User interface of the simulation manager

10.3.5.2   Control features of the simulation manager

The simulation manager (Figure 10.6 shows its GUI) has been created for the convenient on-line modification of simulation parameters that are specific to MoTiV-PTA. Mainly, the request frequency distribution can be modified during the run of a simulation. Furthermore, the simulation speed can be changed for demonstration purposes. The following parameters can be manipulated:

Temporal distribution (1)–(4) The simulation manager allows us to specify a periodic request frequency in order to simulate times of different work load. The experimenter can define a sinus curve by specifying the values of the maxima (1) and the minima (2) of the curve (in number of requests per minute) and its period (3) (in hours). Whenever a request has been posted by one of the UDAs, the world server computes the next time point for posing a new request according to the above specified temporal distribution by computing a value which expresses how long to suppress the emission of a new request. Additionally, a normal distribution is laid on top of the periodic distribution in order to blur the sinus curve: the previously computed delay time is taken as the expected value, while a user-defined factor (4) of that value is taken as the variance.

Spatial distribution (5) A spatial distribution of request can be realised by defining several UDAs to run on differently located servers. The simulation manager allows us to specify on-line a value between 0 and 1 for each of the created UDAs. This value denotes how frequently each UDA launches a request. Again, the actual likelihood for each UDA is computed by dividing this number by the sum of specified numbers.

Type of requests (6) In the MoTiV-PTA demonstrator the user can choose from a variety of different services, such as inter-modal route planning, car park reservation, etc. The experimenter can specify the likelihood of each type of request to occur. This is done by using a slider to assign a number between 0 and 1 to each service. The actual likelihood of this service is then computed as the ratio between the specified number and the sum of all inserted numbers.

Simulation speed-up (7) By using a certain slider the experimenter can speed up the simulation. When put in the middle value, communication delay is reduced to one third while service duration is reduced to 40. Requests are emitted three times faster. When put on the right value, communication delay is even reduced to one tenth while service duration is reduced to one fifth. New requests are then emitted five times faster.

Starting and Stopping the MoTiV-PTA part of the simulation (8) By pressing bottom *start requests*, the UDAs are forced to emit request according to the above frequency distribution. If the *no further requests* button is pushed, the UDA will no longer pose requests; emitted requests are still being answered by the system.

Result Box (9) This box displays the process duration of previously answered requests. This information is then passed to the world server which determines the processing time and stores it for statistical analysis.

According to these specifications, the delay time, the type of request and the UDA are determined. The UDA in question is after the delay triggered to emit the corresponding request. Once a request has been answered by MoTiV-PTA to the UDA, this agent reports the answer to the world server.

10.4   Optimisation of MoTiV-PTA with GRAIL

10.4.1   A generic resource allocation and integration algorithm (GRAIL)

GRAIL (Gerber, 1998) is a multi-staged resource adaptation/meta-reasoning mechanism in which an agent society, agent groups and subgroups are all represented explicitly by *monitor agents* or *representative agents* whose task is to optimise the structure of the represented society, groups or subgroups. This multi-staged meta-level reasoning enables us to overcome complexity explosion.

For each of these monitor agents, we regard the task to adapt the structure of the associated group or society of agents to the current environment as an optimization problem by characterising a *search space* and an *objective function* to be optimized. The *objective function* has to denote the current system's performance while a multi-dimensional search space must describe the system's set of possible configurations. With regard to the application, the objective function has to be defined depending on several factors the application designer has to combine, such as operating time, quality of the result, etc. Each modifiable property of the system reflects one dimension of the search space.

As global optimality can hardly be achieved in a reasonable amount of time, but may be lost very easily, we incorporate a mechanism for achieving and maintaining relatively high performance during the complete run of an application. This mechanism is based on the local search method for finding local optima. A detailed description can be found in Gerber and Jung (1998).

10.4.2   Overview of the approach

The question of how to achieve scalability and stability of the complete MoTiV-PTA system depends heavily on the accuracy of the server configuration. In order to meet these goals, we propose to integrate an optimisation scheme for the server society for the off-line case as well as for the on-line case.

In the following, we exemplify such an optimisation by using the simulation environment of the previous section. This simulation engine is based on a demonstrator of MoTiV-PTA and hence simplifies circumstances in order to show in-principle applicability. The optimisation presented now is based on this system; therefore its purpose is to show feasibility of the GRAIL approach.

In the simulation scenario we assign the quantities *failure probability*, *number of channels* and *maximum speed* to each server. The goal of the optimisation scheme is to find optimal configurations for all servers, i.e. optimal trade-offs between the three quantities. In the on-line case, these optimal trade-offs have to be modified if the environment changes (e.g. the frequency distributions of type of requests, their origin may vary).

In order to model such a trade-off, we introduce the notion of *abstract currency*. In an off-line simulation abstract currency can be used to simulate the payment of fixed hardware costs; in an on-line case, credits of abstract currency can additionally be used to pay for costs of running the system (such as personnel or hardware replacement costs). A fast machine is more expensive than a slower one and a machine with smaller failure probability or one with more UDA channels is more expensive than a less reliable one or one with only few channels.

Besides the optimisation of each server configuration, there are macro-level issues to be regarded, such as the *number of servers*, their *geographical dispersion* or topology, the *distribution of abstract currency credits* to the various servers given a fixed overall maximum, etc.

Clearly, for the relatively independent tasks of server optimisation, a *decentralised problem solving* approach is suitable. However, there are also arguments promoting a *centralised approach*:

- The above macro-level issues favour a centralised decision making procedure.
- The system has to be optimized according to the global task to minimise the average run of a MoTiV-PTA service, as stated above. In addition, feedback is also returned in a global fashion: the actual duration of the runs.

Since both centralised and decentralised approaches have their pros and cons, we employ the GRAIL approach, which unites the two perspectives. It is instantiated on two stages: a global stage and an underlying local stage. Running GRAIL on the server agent society allows dynamic adaptation of each single server and the entire server topology to the current demands posed by the environment. This technique is related to agent-based optimisation of ATM networks (see e.g. Hayzelden, 1998); the main difference is that our optimisation operates on the servers (i.e., the nodes in the network) while ATM optimisation considers the connections (i.e. the arcs in the net). Optimising connections is not applicable in our case, since the MoTiV-PTA system runs on the Internet, which cannot be manipulated in such a fashion.

For each server, we can run GRAIL to optimise its configuration: *failure probability*, *maximal speed* and *number of channels* are now considered as dimensions in the local search space; it is the aim to find an optimal configuration given a fixed amount of abstract currency. Therefore, we now use SIF agents with self-adapting functionalities to represent servers and to perform local optimisations. Server agents are extended to additionally control an instance variable *abstractCurrency* which stores the amount of assigned credits.

We also introduce a higher-level SIF *monitor agent* for the global optimisation view. This agent adapts the system to the current state on the societal level: It tries to optimise the allocation of macro-level configuration, which depends on the *overall number of server agents*, their (logical) *positions*, the *minimum distance of UA migration*, but also the distribution of *abstract currency* to the server agents.

10.4.3   Implementing the optimisation procedure

According to the GRAIL paradigm, each server agent receives *guidelines* from the monitor agent and reports on its current state as a *local profile*. In this scenario, guidelines are implemented through the abstract currency distributed by the monitor agent to the members. A local profile of a member agent contains the number of communication acts of MoTiV-PTA agents logged on that server, the number of connections of MoTiV-PTA agents to that server and the number of services running on the server. A server agent optimises its configuration according to the average run time per service performed on this server. All these optimisation processes run in parallel in order to provide a coherent adaptation of the whole system to changes in the user behaviour. SIF agents now actively take part in the optimisation process. Hence, server agents extend SIF member agents by the simulation functionalities, as shown in Figure 10.7. In this scenario, the monitor agent has no additional domain-oriented functionalities.

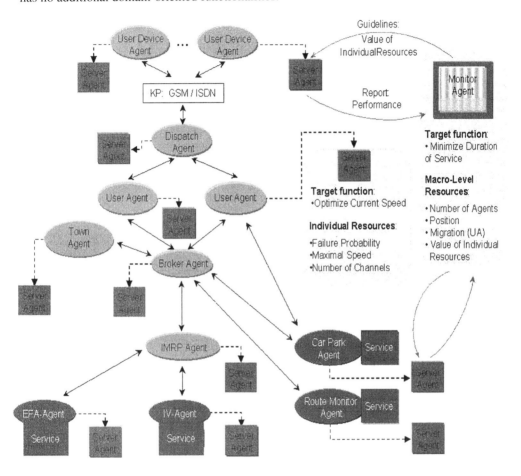

Figure 10.7 Server optimisation in the MoTiV-PTA domain

10.4.3.1   The micro-level optimisation

At the micro level of each server, the abstract currency assigned by the monitor agent has to be spent for the following dimensions:

- *Maximum speed*: We define an abstract measure of speed with a range between 0 and 400 on an integer-valued cardinal scale.
- *Number of channels*: (Range between 0 and 50 on an integer-valued and cardinal scale.
- *Failure probability:* (Range between 0 and 1 on a continuous cardinal scale).

The amount of abstract currency (AC) is computed according to the following formula:

$$AC = \left[\frac{maxSpeed}{500} + \frac{numberOfChannels}{50} + (1 - 4 \times failureProbability) \right] \times 100$$

The standard step size for shifting AC credits from one micro-level dimension to another is 5 AC units. The objective function is derived from the throughput of the server: it is desired to achieve per time unit a high number of communication acts of MoTiV-PTA agents that are located on the servers. A low number may have two reasons: either the server is too idle in the sense that there are too few MoTiV-PTA agents logged on, hence hardware resources are wasted; or the server is too overloaded in the sense that the communication activity of the MoTiV-PTA agents logged on the server is delayed. Both cases are disadvantageous. Therefore we measure a server's performance by *number of communication acts of logged-on MoTiV-PTA agents per minute*.

10.4.3.2   The macro-level optimization

We consider the following dimensions at the macro level:

- *Number of servers:* (Range between 1 and 100 on an integer-valued cardinal scale).
- *Right to change the minimum migration distance*: The distance between two servers on the grid can be at most 30 units (on an integer-valued cardinal scale).
- *Abstract currency*: Servers may have between 30 and 300 credits at their disposal; to break complexity, a modification on that dimension is modelled as an allocation shift from one server to another due to the assumption that it is always reasonable for a server agent as well as for the entire society to spend all of its abstract currency and hence an increase of spent units must be compensated by an appropriate decrease elsewhere.

Overall performance is measured by averaging the *duration time of a representative variety of MoTiV-PTA services over a time period*.

Changing the number of servers of course influences the server topology on a map. Modifying this quantity hence amounts to changing the server structure. We use the following heuristics to decide where to place an additional server or which server to shut down.

Putting up a new server

Whenever the monitor agent decides to increase the number of servers during the run of a macro-level optimisation, it has to be determined where exactly to place this server. Two sometimes conflicting criteria have to be considered:

- *Distance between servers*: In order to minimise communication delay, the server topology should provide a uniform density, i.e. servers should be spread with almost equal distances over the map, considering the shape of the boundaries of the map.

- *Work load of servers*: Overloaded workstations need support, hence it is reasonable to place new workstations closer to stations that are bottlenecks.

Strategies for finding optimal solutions to this problem have been proposed in the field of OR (see for instance Juel and Lowe (1985) for an overview). However, they usually run in an iterative way at rather high computational costs. For the purpose of demonstrating applicability of the optimisation approach, we use a heuristic-based technique to find a position for a new server. It is not required to find the perfect position but an acceptable one without spending too much computational time and space. The monitor agent maintains eight *Candpoints* for each server agents: Each point is the exact intermediate point between the position of a server and the boundary of the map according to the four cardinal points of the compass and their intermediate directions NW, SW, SE, and NE. If bottlenecks are not considered, the position of a new server will be assigned to that point which maximises the maximum distance to all existing servers. (This strategy is intended to meet the requirement of a relatively equally distributed topology.) Finding such a server is performed in the background when a new server is introduced or a server is deleted.

In order to address the second requirement, the current status of each server has to be regarded. Hence, whenever a new optimisation step is performed, the work load of each server α_i is taken as a certain factor a_i by which the candidate position is narrowed to the server: servers with more than average MoTiV-PTA agents and less than average performance favour a closer position of the new server. The factor a_i is calculated according to the following function:

$$a_i = \min\left(1, \frac{\text{performance}(s_i)}{\text{performance}_{avg}} \times \frac{\text{PTAAgentlist}_i}{\text{PTAAgentList}_{avg}}\right)$$

The *new* candidate position is then assigned to $a \times pos(s_i) + (1-a) \times pos(candPoint)$. Since the performance of a server changes during the run of the system, these factors are computed for all servers during each optimisation step; so is the determination of the one point that maximises the maximum distance to a server in the system.

The new server agent of course needs abstract currency units for its local optimization. Since the overall number of abstract currency credits is limited, the new server agent earns 30 units that are removed from other agents in equal parts.

Shutting down a server
If the monitor agent decides during an optimisation step to decrease the number of agents, a server to be closed down has to be found. As a simple heuristic we select the one server with minimal abstract currency credits, since this server is supposed to have only little contribution to the overall performance. Its abstract currency units are distributed equally to the other agents.

Before a server agent is terminated, all MoTiV-PTA agents that are currently logged on that server are forced to migrate. New servers are selected according to the previously stated trade-off between distance and server speed.

10.4.4   Integrating heuristics from bottleneck analysis

We can integrate knowledge on the internal state of a server agent into the decision-making process. Besides finding a location for a new server, server workload is also used to specify the shift of abstract currency: low-performing servers (i.e. servers with relatively few communication acts of logged MoTiV-PTA agents) but with *many* logged MoTiV-PTA agents and services are most likely overloaded; hence shifting currency units to such an agent is expected to increase not only its local performance, but also global performance.

On the other hand, servers with few communication acts of logged MoTiV-PTA agents and *few* logged MoTiV-PTA agents and services are assumed to be too idle: hence resources are taken from such an agent and assigned elsewhere. In this case, the performance of the agent will clearly further drop, but global performance can be expected to increase.[2]

In the heuristic-based optimisation scheme for the MoTiV-PTA application the above observations are incorporated as follows. If the number of connected MoTiV-PTA agents that are connected to a certain server exceeds the doubled average number of MoTiV-PTA agents or services, then this server is kept in memory. If several servers fulfil this requirement, the one with the highest resources is taken. If there exists another server which a) is less than half loaded than an average server and b) has minimal performance (if there exist several ones) then the former server then receives credits from the latter. The amount of transferred credits is not determined by this heuristic, but by the main optimisation procedure.

If the amount of assigned credits of a certain server has been dropped to the minimum value (which is set to 30 in this case) while the overall performance keeps on increasing, that particular server is deleted and its remaining resources are distributed to other servers. If the number of connected MoTiV-PTA agents connected to a server is less than half of the average number of MoTiV-PTA agents or services, then this server loses credits from a server with a) a more than double of the average workload and b) has minimal performance (if there exist more than one).

10.4.5   Integrating heuristics from machine learning

In this scenario, machine learning techniques are integrated to find a promising direction and size for a step in the search space: a *memory base* module stores situations and actions which have led to significant performance decreases or improvements. The key idea is to remember such a situation later in a new situation and to take the previously performed significant action as a suggestion for a future action.

We store situation action tuples in a hash table. To cut down the complexity of finding appropriately similar situations later on, we introduce *profiles* which are similarity classes of the situations. Since the number of possible profiles is still greater than the number of possible actions in this scenario, it is more efficient to store profile-action tuples by using

---

[2] This resource shift is possible since GRAIL member agents are benevolent to their monitor agent and accept guidelines.

the action as the key. Two profiles are merged to one if an action has proven efficient in both situation classes.

Whenever a new step direction has to be determined, the current situation is matched against the memory base by parsing through all situation profiles. If a stored profile is found which includes the current situation, a corresponding action is performed if the former action had led to a performance increase. If the stored action had decreased the performance, an action in the opposite direction is done.

For long-term applications it is important to keep the memory base at a size that allows a fast situation matching. Since we assume a continuously changing environment, older experiences may be outdated and hence have to be removed from the memory base to keep it small enough. Hence we annotate each entry with a weight which is decreased over time. Whenever a profile has led to some good advice, its weight is increased, otherwise it is even further decreased. Once the weight has reached 0, the corresponding profile is removed from the memory base.

One of the major advantages of this approach is that there is no explicit need for a training phase. However, this mechanism can only provide good advice if the memory base is adequately filled. Therefore, the system must have run for a while before good results can be expected.

10.4.6   The Extended User Interface

Figure 10.8 shows a screen shot of the complete system: the MoTiV-PTA agents DF and AMS are depicted on the very left; the figure displays the scenario GUI and the simulation manager as well as the object info windows of the monitor agent (third window on the left) and one server agent (the rightmost window).

10.4.6.1 Additional control features of the simulation manager

For optimisation purposes the simulation manager offers facilities to modify the type of optimisation while running the optimisation. The experimenter can choose for macro-level optimisation between no optimisation, a greedy strategy, a greedy strategy with integrated machine-learning optimisations, a greedy strategy with integrated bottleneck analysis or one with both machine learning and bottleneck analysis optimisations. For micro-level optimisation, they can turn on or off a greedy strategy. More details on these strategies are provided in Gerber, Steiner and Bauer (1999).

10.4.6.2 Extensions of the SIF script parser

For running the optimisation, the simulation script parser needs only to be extended to allow the introduction of a monitor agent:

MONITORAGENT *<name>, <xPos>, <yPos>* A script entry beginning with the keyword MONITORAGENT leads to the creation of a SIFIRA monitor agent named <name>, located in the map at position *[<xPos>;<yPos>]*. Although this agent does not correspond to any real-world entity, we place it on an unused point of the map for visibility reasons.

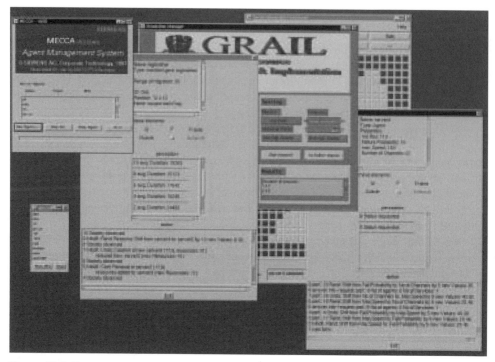

Figure 10.8 Screen shot of the entire system

10.5   Conclusion and Future Work

In this chapter we have presented a case study that employs the GRAIL techniques and makes use of the SIF development kit. In this case study we have extended the MoTiV-PTA demonstrator to a simulation engine and we have integrated a self-adaptation mechanism. Future work may focus on further refining the simulation engine and the optimisation scheme.

Acknowledgements

We would like to thank Donald Steiner and Bernhard Bauer from Siemens AG for providing the MoTiV-PTA demonstrator.

10.6   References

Funk, P., Gerber, C., Lind, J. and Schillo, M. (1998) SIF: An Agent-Based Simulation ToolBox using the EMS Paradigm. In *Proceedings of the 3rd International Congress of the Federation of EUROpean SIMulation Societies (EuroSim), 1998.*

Gerber, C. (1998) Bottleneck Analysis for Self-Adapting Multi-Agent Societies. In *Proceedings of the IEEE Joint Conference on the Science and Technology of Intelligent Systems, 1998.*

Gerber, C. and Jung, C. (1998) Resource Management for Boundedly Optimal Agent Societies. In *Proceedings of the ECAI Workshop on Monitoring and Control of Real-Time Intelligent Systems, 1998.*

Gerber, C. Steiner, D. and Bauer, B. (1999) Resource Adaptation for a Scalable Agent Society in the MoTiV-PTA Domain. In A. Hayzelden and J. Bigham (Eds), *Software Agents for Future Communication Systems.* Springer Verlag, 1999.

Hayzelden, A. (1998) Telecommunications Multi-Agent Control System (Tele-MACS). In *Proceedings of the European Conference on Artificial Intelligence (ECAI), 1998.*

Juel, H. and Lowe, R. (1985) The Facility Location Problem for Hyper-rectilinear Distances. *IEEE Transactions,* **17**, pp. 94 -98.

Russell, S. and Norvig, P. (1995) *Artificial Intelligence: A Modern Approach.* Prentice Hall.

Schillo, M., Lind, J. Gerber, C. Funk, P. and Jung, C. (1999) SIF - The Social Interaction Framework System Description and User's Guide to a Multi-Agent System Testbed. Research Report RR-99-02, German Research Center for Artificial Intelligence.

11

Flexible Decentralised Control of Connection Admission

J. Bigham, L. G. Cuthbert, A. L. G. Hayzelden, Z. Luo

11.1   Introduction

This chapter describes architectural aspects of a distributed multi-agent system that manages admission to a telecommunications network and attempts to demonstrate that the system constructed provides more flexible management options than are available in conventional signalling systems. Flexibility is important, as future telecommunications systems will support many more kinds of service and more 'intelligence' than that offered to users today and users will expect to have choices of provider and qualities of provision (Hayzelden and Bigham, 1999a).

Because the agent system described spans the network management plane and the network control plane, several reactive layers and a tactical planning layer are used to control admission. Ideas have been taken from the subsumption architecture (Brooks, 1986) and, as such, the agent system shares the feature that if a higher layer is not operational then the system still operates, albeit in a degraded fashion.

There are many possible future business models. We have chosen one to illustrate the approach in the management plane such that a network provider (NP) manages an entire physical network. The NP may lease the use of bandwidth to some service providers (SPs). The NP also carries services and so is an SP in its own right, but it will have more management options as it owns the physical network and so knows more about its topology and utilisation. The SP owned by the NP will be denoted as the network service provider (NSP). The SPs are in competition and when a customer requests connection to the network it has a choice of different service providers. To the NP, the NSP is a cooperative agent, whilst the other SPs can be viewed as self-interested agents. In this chapter we do not cover strategic inter-SP and SP-with-NP negotiations. Rather, we assume, for example, that the bandwidths leased and the logical topologies provided to the SPs have been decided, and we are concerned with the effective management of an SP's network or the NP's network at the tactical and control level. The architecture described supports customer choice at entry to the network and provides mechanisms for flexible real-time bidding and connection management for both service and network providers. The approach is applicable to any type of network where there is a connection set up procedure.

Since a network is a distributed system, network control involves an interaction protocol, called the signalling protocol, for exchanging control information (connection requests, topology updates) among the network elements. However, there are problems interfacing agents with network devices. Interfacing agents with the network requires some form of programmable interface to the network devices as it is necessary, for example, to override the conventional ATM signalling. At the moment, programmable network device interfaces are not always available, particularly in commercially available devices, though standards are being agreed (Biswas et al., 1998). To solve this problem generic operations have been implemented using management protocols. An architecture for embedding non-proprietary network control and resource management functionality into ATM networks has also been developed, but that architecture is described elsewhere (Vayias, Soldatos, Bigham, Cuthbert and Luo, 1999).

A feature of the architecture is the integration of the reactive components into the bidder of the SP agents and the ability to 'plug-in' different policies for these reactive components. Different policies are appropriate for NPs and SPs. The bidder is the central mechanism that allows competing SPs to sell deliverable quality of service (QoS) to customers at different levels of customer demand. The agent acting on behalf of the user can modify the behaviour of the auctioning process to allow for anticipated demand.

All the mechanisms have been implemented and successfully controlled admission on a real experimental telecommunications network. Results from experiments and performance issues resulting from management protocols are discussed in Vayias et al. (1999).

11.2   The Agent Model

The control and resource management system has six kinds of agent, namely, the Service Provider Agent (SPA), the Network Provider Agent (NPA), the Proxy User Agent (PUA), the Connection Agent (CA), the Resource Agent (RA) and the Switch Wrapper Agent (SwWA).

Since the testbed used to validate the work is an ATM network, the description and implementation given here are based around ATM. Similar quality-of-service problems are now being addressed in the IP domain – much of what is described is applicable to both protocols. ATM is a connection-oriented transfer mode offering a guaranteed QoS between a source and a destination. A source and destination are called a *source destination pair*. We assume for definiteness that the QoS is provided through a choice of one of three classes of service (CoS) and the specification of suitable traffic parameter values. QoS parameter values are specified for both the forward and backward directions of the connection within the chosen class. It is also assumed here that the network resources will be managed by exploiting dynamic bandwidth allocation to *virtual paths* (Cuthbert and Sapanel, 1993). A virtual path (VP) is an acyclic path of specified bandwidth from a source node in the network to a destination node in the network using physical links of the network (Friesen, Harms and Wong, 1996). Exactly how the bandwidth is specified depends on the CoS chosen (Vayias et al., 1999). Note that only source-to-destination virtual paths are managed. All new connections are allocated to one of the VPs. The bandwidth associated with any VP can change continually and is one of the controllable parameters for an SP or NP. It is assumed that the set of VPs associated with a source destination pair are known – fixed in terms of route though not bandwidth – and are a manageable subset of the set of possible VPs for that source destination pair. In practice, the set of enumerated VPs could be changed over time. In Figure 11.1, the NP is represented by the Network Provider Agent (NPA). The competing service providers are represented by their Service Provider Agents (SPAs). Each SP is illustrated by a different plane in the diagram. The NP also carries services and is also a plane in the diagram. The top plane shows three VPs for a particular source destination pair and a particular CoS.

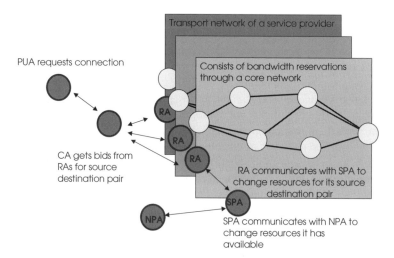

Figure 11.1   The different SP networks viewed for a single source destination pair

The model described is particularly appropriate for a service provider since it cannot change the route of a VP without arrangement with the NP. (The switch cross-connects of a VP are set up by the NP, via the SwWAs.) An SP may not even know the physical paths, though some knowledge of the topology of the SP's leased VPs is assumed for better decision making at high utilisation levels. The knowledge used is only that which can be deduced from the terms of the lease agreements with the NP. However, an SP can readily dynamically modify bandwidth allocation to its VPs, so long as it keeps within the capacity limits agreed with the leasing NP. Each SP owns an RA for each source destination pair it services, and this RA manages the resources of the VPs that belong to the source destination pair. So with multiple service providers and multiple source destination pairs, there are many RAs. The user requests a connection of a particular CoS along with the associated CoS parameter values via the PUA. The PUA contacts the CA placed at the entry point to the core network. The CA conducts an auction on behalf of the user, the bidders being the RAs of the different SPs that service the source destination pair.

One issue that is obviously of interest with real networks (but harder to demonstrate on the testbed) is the problem of scalability. The architecture considered addresses this issue because it allows for networks to be partitioned into domains and a modified form of the

PUA, called a Proxy Connection Agent (PCA), could be used to manage connections between networks instead of connections at user equipment.

The CA queries the RAs for bids and gets the replies from relevant RAs; it then decides on the preferred service provider, or the preferred offer from that service provider in the case of multiple offers, and instructs the chosen RA to install the connection. The chosen RA then interacts with the Switch Wrappers. The SwWA constitutes a 'virtual' software abstraction of an ATM switch and its resources, and provides a generic, vendor-independent interface for network control and management applications. Interaction between the RA and SwWAs is performed by means of an agent communication language (FIPA, 1997), while the SwWAs have access to the switch's proprietary control method (e.g., via simple network management protocol (SNMP) or some form of API). Figure 11.2 shows the physical locations of the agents.

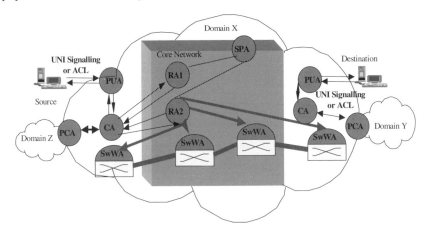

Figure 11.2   Location of the main agents in the IMPACT system

11.3   Competing Service Providers

Part of the flexibility offered by the architecture is its ability to support the selection of alternative service providers by customers. The user, through its PUA, chooses an SP by a modified first-price sealed-bid auction. SPs can enter and leave the market dynamically and could even be unknown to the customer.

The auction is a mechanism for letting the user know if the SP is prepared to offer the requested bandwidth at the requested QoS and for the user to obtain a price, if there is no pre-negotiated tariff structure. So whilst the bid price is derivable by the user when there is a tariff structure, the intention to accept the connection request is not. The heterogeneity of the market and the need for the market to be open, makes the mechanism of an auction appropriate.

The auction is modified to allow the user to use the tariff structures to make comparisons with other tariffs and one-off bids. For example, suppose that when a user buys more than a threshold quantity from an SP, a discount is given. Then this user can use its private knowledge of its expected requirements in the current accounting period to decide whether it is worth being loyal to a particular SP. The PUA passes instructions to the CA as a list of (SP Id, adjusted price) pairs for those SPs for which negotiated tariffs exist.

Example. Assume for simplicity that all connection requests are for the same amount of bandwidth. Suppose a customer has a contract with SPA1, who according to the negotiated agreement will reduce the price per unit from 25 to 5 in stages, when threshold volumes are reached in the accounting period (see Figure 11.3). The customer has no contract with SP2, which bids 15, and there is no reason to believe SP2 will change its behaviour. Suppose the customer has already bought 4000 units and expects to use 30000 units in the financial period, then its expected price for a unit of bandwidth, if it is loyal to SPA1, in the financial period is 10.965. Therefore, the list of adjustments passed to the awarder is {(SPA1, 10.965)}. The CA simply executes a first-price sealed-bid auction on the modified prices, so SPA1 would win the auction.

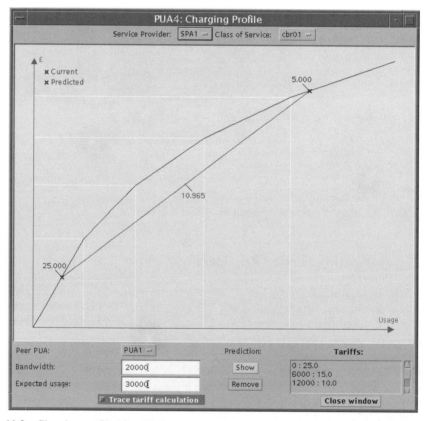

Figure 11.3 Charging profile of an SP for a particular customer

The more complex the negotiated terms, the more complex the nature of the look-ahead required by the PUA. Our architecture leaves such intelligence with the user, or with the

PUA, and not with the CA. Whatever the look-ahead problems, it is assumed that the calculations map into a set of adjustments for the awarder (described below) in the CA.

The auction has four components: *an announcer, a bidder, an awarder, and an award taker* (Sandholm, 1999). In the context of a CA and SP auction, we are only concerned with the announcer and awarder of the CA, and the bidder and award taker of the RA. When a CA agent receives a request for bandwidth from A to B from the PUA, its announcer makes a call for proposals, i.e. an announcement, to those RAs that manage the resources of SPs that have VPCs from A to B. The SP has provided each of its RAs with a bidder and a reward taker. The CA agent is conceptually located at A, and is probably, though not necessarily, physically located there. The simplest assumption is that bids are taken as binding and so if awarded the connection must be made and the offered QoS provided. In the implementation, a great deal of effort has been spent on trying to achieve this. This involves reserving space to provide the required QoS at the time of bidding, and releasing all bandwidth reserved if the bid is unsuccessful. In practice, such reservations by all bidders are very wasteful and almost certainly such a high degree of assurance of QoS at the time of the bid is not necessary for most users. However, if reneging is allowed, there is a temptation for an SP to bid more than it should, in the unreasonable hope that space becomes available, often wasting the customer's time. Unchecked, an SP could indulge in even more selfish behaviour and renege on a bid deliberately, if a more profitable connection request announcement is subsequently received. If reneging is allowed for ordinary connections (i.e. the majority that do not require capacity transfer or connection transfer), then more sophisticated protocols are needed which include penalties. These could be penalties generated using statistics collected by monitoring regulatory bodies (such as Oftel in the UK), or more weakly through self-regulation driven by poor customer perception.

11.4   Resource Management Strategies

Apart from providing a mechanism to let the user negotiate access to the network, the different service providers have to manage the network resources to satisfy physical constraints and meet QoS parameters. This section describes the approach taken to resource management.

For clarity we will call an SP that is not an NP a secondary service provider (SSP) and the SP that is the NP the NSP, and use 'SP' if we mean either. The NSP and each SSP have planning capabilities. We will only consider capacity bandwidth allocation. The planning levels of the NSP or SSPs are informed about bandwidth usage from the RAs they own, so that they can build a picture of the usage of the network they own or rent. The capacity planner works in the management plane. The planning suppresses the view of the RAs' world by altering the capacities that they work in (Hayzelden and Bigham, 1999b; Bigham et al. 1999). The relationship between the NPs, SPs and RAs is shown in Figure 11.4.

The planning for an SSP and for the NSP are slightly different. Only the former will be described. The SSP planner uses source destination pair demand, not individual VP demand, though we do use existing VP demand as a constraint in the planner to avoid producing an infeasible plan. (Some care needs to be taken when a reduction is requested. We cannot reduce the capacity of a VPC to a level below that which can be attained from the disconnection rate, based on the current usage of the VPC. The planner will have to move in the optimal direction, and possibly move further the next time.) The 'link' capacities purchased from the NP are also input. These links need not be true links in the physical network but are links in the SSP's model of the network. The planner then computes an allocation of VPC capacities that maximises a function of the minimum residual capacity in the links of the network, breaking ties using hop count. The planning level sets capacities for the reactive allocation policies, i.e. gives each RA a bandwidth allocation for each VP managed by that RA. The allocated bandwidths are scaled so that the minimum residual of the capacity allocations over the network is zero. So the bandwidth capacities of the VPCs are changed periodically. The optimisation criterion has the property that if source-destination demands are scaled pro rata, up or down, over all source destination pairs in the network, then the ratios of the capacities computed are still optimal. This means that the planner can always find an optimal allocation where the minimum link residual is zero and give the distributed RAs as much capacity as possible to play with.

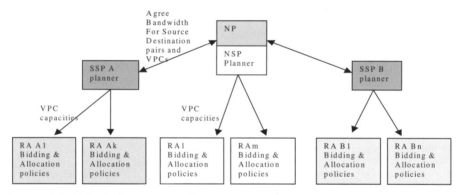

Figure 11.4   Principal competence layers

We now describe the architecture of the RA. Its main components are shown in Figure 11.5. There are three reactive layers associated with the bidder. The lowest *check-capacity* layer handles bidding using the capacities provided by its owner's capacity planner. The two higher layers, called *capacity transfer* and *call transfer*, respond to unexpected demands and make adjustments to the previously 'optimal' allocation provided by the planner. These are necessary as the planner will not always correctly anticipate the demand profile that will be experienced across the network. The higher layers in the bidder allow runtime correction to the plan; these corrections are cumulative and hold until the next re-planning. This avoids unnecessary call rejection. The check-capacity layer of an RA works using two strategies: a selfless strategy, which we will call *proportional allocation*, and a selfish strategy which we will call the *fragmentation reduction*. Work on societies of agents, e.g. Hogg and Jennings (1999), has shown that the performance of a society can change as the proportion of selfless and selfish agents changes. For an SP, we base the transition from the selfless to the selfish policy on the utilisation of the source destination pair. At low-to-normal utilisation, we adopt proportional allocation; at high utilisation, we adopt fragmentation reduction.

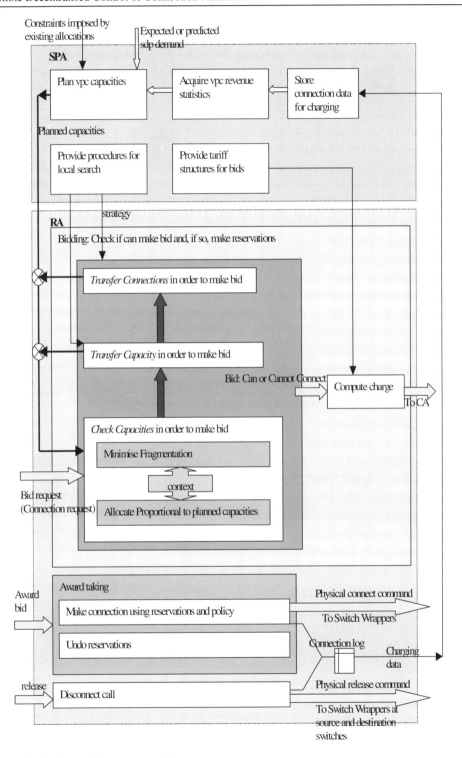

Figure 11.5   The architecture of an RA

When a bid cannot be made, capacity transfer can be considered. This process is made fast and local to the bidding RA by keeping, for each RA, a pre-computed table of covering sets for each VP of the RA, in preference order. This table is local to the RA, and the bidder performs all actions based on the information in this table. The SP can update this table if new capacity transfer strategies are to be used. (In the implementation this step is not done.) The covering sets for a VP are tried in preference order and if none of the covering sets has the capacity needed the bidder does not bid. If capacity transfer is performed, the SP is informed. Figure 11.6 illustrates a simple example.

Example. RA2 has only one VP currently carrying 8 units, 0 reservations and a capacity of 10. A request for 4 units of bandwidth is received by RA2's bidder, which it cannot provide. It can only provide 2. It reserves 2 units. All reservations are undone if no capacity transfer is possible, or if the bid fails. RA2 then searches for VPs belonging to RAs owned by the same SP for a set of VPs, which together cover the VP in question. Here two sets can be found, each containing one element. VP b and VP c of RA1 both cover the VP of RA2. The bidder picks b first, say, and checks whether two units of capacity can be transferred. (In fact c would be the smallest superset as it has fewer links not belonging to the VP in question.) Since it can, the final reservation of 2 more units is made here. If the bid is refused, the reservations and capacity transfer are undone. If the bid is successful the capacity transfer is kept, and the reserved bandwidth is changed to used bandwidth.

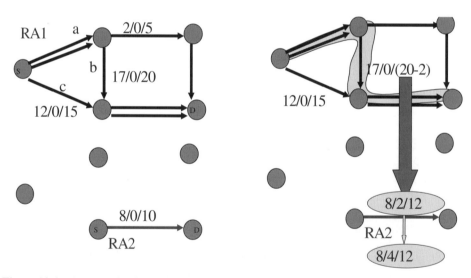

Figure 11.6   An example of capacity transfer

Connection transfer builds directly on capacity transfer, and is appealed to, when the user permits, when capacity transfer on its own fails. Because these are real connections it may not be possible to transfer the *exact* amount required, so we transfer at least the amount required. However, there could have been several disconnections since the bidder assessed the situation and there may not be the required quantity of transferable bandwidth still connected. In the current implementation we take the view that the connection transfer option does not offer a 100% chance of delivering the required QoS with a high probability, though in principle this could be done.

11.5   Conclusion

This chapter has concentrated on describing the architecture of a multi-agent system which performs traffic control and resource management. The agent system works in a distributed and decentralised manner. Only the planning layers are centralised in their associated SPs. Many important details related to signalling, the details of the switch wrappers, the mechanisms performed within the reactive layers, and the process of establishing the physical connections have been omitted. The features that distinguish this approach from the conventional method include:

1. The admission decisions are taken at the access node, thus alleviating the burden of executing the conventional conncetion admission control procedure in every individual switching node, giving fast response times.
2. The agents are structured in competence layers, and work at different time scales, which allows a balance of reactive and planned behaviour and resilience to control system failure.
3. The components have many of the characteristics of agents; for example:
   - Each RA works independently based on its view of the world passed down by the SP or NP giving autonomy
   - Each RA can change from cooperative to self-interested depending on its local state giving adaptable social ability
   - The different layers work at different speeds, the lowest three layers being fast enough to make a connection in real time, giving responsiveness
   - In some circumstances, an RA can change not only the configuration as passed down by the planner, but also the configuration of other RAs owned by the SP, in order to make a bid, i.e. it exhibits pro-activeness.
4. The architecture supports competing service providers and a competing network provider within a realistic open bidding structure.

Although an ATM network is used as the basic infrastructure, many of the concepts are generic and the IMPACT system can handle IP traffic equally well. In fact, the overall concept is of a fairly dumb ATM core network with management plane switching (cross-connects) and intelligent edge switches controlling resources. Such a network represents a structure used by operators around the world.

The prototype of the agent system has been designed and implemented. Sets of trial scenarios have been evaluated on an ATM testbed and are reported in Vayias et al. (1999). These experiments, on commercially available switches, have confirmed the practical feasibility of exploiting the flexibility offered by the different layers described when making real-time connections.

The IMPACT project has implemented a streamlined, purpose-designed agent software system called basic agent template (BAT). BAT provides a framework that satisfies the agent technology requirements as well as the special real-time requirements imposed by the IMPACT project itself. Communication is achieved by the exchange of Java objects using a subset of FIPA communication acts and, thanks to Java's Reflection Mechanism, agents determine at runtime which method is to be used to handle a received object.

Acknowledgements

Much of the work described in this chapter was part of the ACTS IMPACT project. The authors gratefully acknowledge support from the European Commission under the ACTS Project AC324. The authors also acknowledge valuable help and contributions from its partners Swisscom (the authors of the BAT system), TeleDanmark, National Technical University of Athens, Flextel, Teltec, and ASAP.

11.6   References

Bigham, J., Cuthbert, L.G., Hayzelden, A.L.G. and Luo, Z. (1999) Agent Interaction for Network Resource Management. *ICON – Interoperable Communication Networks*, **2(1)**. Available: http://www.baltzer.nl/ [13 June 2000].

Biswas, J. et al. (1998) The IEEE P1520 Standards Initiative for Programmable Network Interfaces. *IEEE Communications Magazine*, **October**, 64–71.

Brooks, R. (1986) A Robust Layered Control System For A Mobile Robot. *IEEE Journal of Robotics and Automation.* **2(1)**, 14–23.

Cuthbert, L.G. and Sapanel, J.C. (1993) ATM: The Broadband Telecommunications Solution, in *IEE Telecommunications Series 29.*

Foundation for Intelligent Physical Agents. (1997) Agent Communication Language. FIPA 97 Specification. Part 2. FIPA, Geneva, Switzerland.

Friesen, V.J., Harms J.J. and Wong, J.W. (1996) Resource Management with Virtual Paths in ATM Networks. *IEEE Network*, **September/October issue,** 10–20.

Hayzelden, A.L.G. and Bigham, J. (Eds.) (1999a) Software Agents for Future Communications Systems. Springer-Verlag.

Hayzelden, A.L.G. and Bigham, J. (1999b) Agent Technology in Communications Systems: An Overview. *The Knowledge Engineering Review*, **14(3)**, 1–35.

Hogg, L. and Jennings, N.R. (1999) Variable Sociability in Agent-Based Decision Making, in *Proceedings of the 6th International Workshop on Agent Theories, Architectures and Languages (ATAL-99)*, pp. 305–318, Orlando, FL.

Sandholm, T.S. (1999) Distributed Rational Decision Making, in G. Weiss (Ed.) *Multiagent Systems* pp. 201–258, MIT Press. Cambridge, MA.

Vayias, E., Soldatos, J., Bigham, J., Cuthbert, L.G. and Luo, Z. (1999) Intelligent Agents for ATM Network Control and Resource Management: Experiences and Results from an Implementation on a Network Testbed. *Technical Report of Dept Electronic Eng., Queen Mary, University of London.*

11.7   Further Reading

Courcoubetis, C. (1998) Pricing and Economics of Networks, in *Proceedings of the IEEE InfoCom.* Available: http://www.ics.forth.gr/~courcou/presentations/infocom/presentation/sld001.htm [13 June 2000].

Guerin, R., Ahmadi, H. and Naghshineh, M. (1991) Equivalent Capacity and its Application in High Speed Network, *IEEE Journal on Selected Areas in Communications*, **9(7)**, 968–981.

12

Low-Level Control of Network Elements from an Agent Platform

M. Hansen, P. Jensen, J. Soldatos, E. Vayias

12.1   Introduction

The IMPACT[1] project has implemented a prototype system for controlling ATM networks where control-plane tasks (connection admission control, routing, Virtual Path management) are completely undertaken by agent software. This meant that some means of low-level control of the network elements had to be provided to the agent software platform, bypassing the control logic embedded in the network elements by the manufacturers. On the one hand, conventional 'telephony-oriented' ATM signalling had to be 'worked around'. However, in order to comply with existing terminals, ATM-Forum or ITU signalling still needs to be supported at the user network interface (UNI). This has generated the need for an entity that can map between standard signalling and the agent language. This entity is called the *Proxy User Agent*. On the other hand, agents should have a way of conveying connection control and configuration commands to the actual network elements. This introduced the need for a *Switch Wrapper* entity which 'wraps' an ATM switch with a generic switch-control interface for exchanging control messages in the agent language.

   This chapter presents the design and implementation of these software entities, which form a low-level control layer between the agents incorporating the network control logic and the network devices themselves. Some general architectural concepts for open programmable network control are presented in section 12.2. Then, the Switch Wrapper is presented in section 12.3 and the Proxy User Agent in section 12.4. In order to demonstrate the advantages and flexibility of the agent-based control approach, real applications have

[1] IMPACT (AC324 - Implementation of Agents for CAC on an ATM Testbed) is a research project funded by the European Commission under the 4th Framework Programme.

been called for. Apart from using simple native ATM applications, IP applications have been considered for a more illustrative demonstration. Section 12.5 describes how IP applications on ATM terminals can use the IMPACT system either with standard solutions such as Classical IP over ATM (Laubach, 1994) which requires no extra adaptations at the terminals, or even how such an application could benefit from QoS-enabled VCs setup by using a Terminal Agent. We thus demonstrate how IMPACT's open control approach indulges more applications (and even existing ones) to benefit from ATM's performance and QoS. A complete specification of the subsystem presented in this chapter can be found in IMPACT (1998a,b).

12.2   Architectural Concepts

A very important issue when implementing an agent-based network control system is to devise and implement mechanisms for interfacing the agent software with the target network devices. To this end, it is highly desirable that the target network devices offer some degree of programmability; however, in most devices such interfaces do not exist. In particular, each device (e.g. ATM switch, IP router) comes with a suite of network control schemes which are specified and implemented by the particular vendor. Internal details of the network control strategies are not released and access to the switch control software is not provided. Even in cases where some commercial switches provide access to their control software through an API, this is limited to configuration tasks and does not allow modification of the core components of network control and resource management mechanisms. Below, we present a quite general architecture that allows the implementation of network control and resource management functionality other than the 'default' behaviour offered by the devices. Building on the concepts of this architecture, we have implemented interfaces between the software agents described and the ATM switches used in the scope of IMPACT experiments.

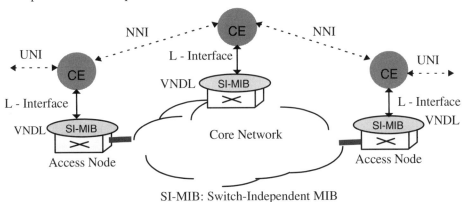

SI-MIB: Switch-Independent MIB

CE: Control Entity (CAC, routing, etc.)

Figure 12.1   Architecture for embedding network control and resource management functionality in ATM networks

Placing an agent-based control system on top of network nodes results in altered control functionality and demands that the default control behaviour of the nodes be overridden. In this context, Figure 12.1 presents an architecture for embedding desired network control functionality in ATM networks which overcomes the lack of full access to the switch control software of most commercial ATM switches. According to this architecture, the network nodes export appropriate well-defined programmable interfaces to distributed software entities that implement various network control schemes on top of the nodes.

The idea of implementing programmable interfaces stems from the recent strong interest in network programmability. Since several companies and laboratories have started a new IEEE standards development project, namely IEEE P1520 (Biswas et al., 1998), it was our belief that the implemented programmability should comply with the proposals and developments of this group. In order to align the architecture depicted in Figure 12.1 to the reference model produced by the P1520 initiative (Figure 12.2), each node should export an L-interface, allowing network control algorithmic components to make use of the so-called Virtual Network Device Layers (VNDLs). A VNDL constitutes a virtual (software) representation of a node, comprising any essential piece of information for the deployment of network control and resource management algorithms. In the proposed architecture, we call the VNDL representation (i.e. the collection of appropriate controlled objects) Switch-Independent Managment Information Base (SI-MIB).

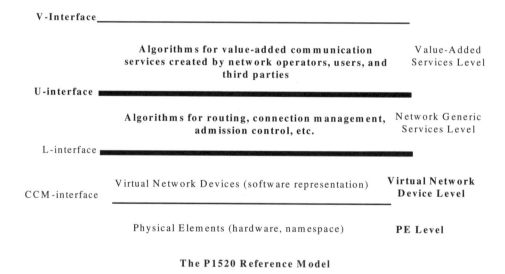

Figure 12.2   Reference model of the P1520 standard for programmable networks

According to our architecture, distributed Control Entities (CEs) deal only with the information contained in the SI-MIBs, which is in accordance with the actual state of the nodes. For the deployment of a network control scheme, the said software entities exchange network control information by consulting their SI-MIB. The interaction between a control entity and its corresponding SI-MIB takes place through a generic L-interface. In order to have a functional L-interface, a mechanism for accessing the low-level capabilities of the actual node is needed. Such a mechanism should enable a control entity to query or alter the status of a network node. This can be performed well through a general-purpose management protocol (e.g. SNMP, GSMP, CMIP), depending on the management capabilities of the target network node. As a result, according to our architecture a network control algorithm invokes some generic operations of the L-interface. The execution of these operations on the actual node takes place through appropriate drivers that interact with the node using a management protocol (in P1520 terms, this is called the Configuration Control and Management – CCM Interface). However, it must be noted that the use of a management protocol (such as SNMP) for controlling the node through a software program residing outside it introduces a performance penalty. In general, management-based techniques are very slow when it comes to performing conventional control operations, which would normally be carried out through signalling.

It is important to note that this architecture allows the network control functionality to be applied on a multi-vendor network. This is because the algorithms actually interact with the SI-MIB, which is vendor-independent. Therefore, the L-interface can also be generic and uncoupled from vendor-specific details. However, each different switch requires the implementation of a new software driver for accessing the low-level capabilities of the device. This inevitably implies some additional effort for installing and adapting the described software system to a multi-vendor environment.

This fairly general architecture for interfacing software control entities to the network, has been used by the authors (apart from IMPACT), in other applications (Vayias, Soldatos, Kormentzas, Mitrou and Kontovassilis, 1998; Soldatos, Kormentzas, Vayias, Kontovasilis and Mitrou, 1999).

12.3 Interaction with the Network Elements – the Switch Wrapper Agent

The Switch Agent Wrapper (SwWA for short) is a software component which is required in order to provide agent 'wrapping' to ATM switches, i.e. to enable agents in a multi-agent system (like the IMPACT system) to retrieve information from and/or control ATM switches by means of a higher-level language than the standard switch management and control protocols (such as SNMP, GSMP, etc.). The SwWA constitutes an abstract software representation of an ATM switch's resources and provides a generic, vendor-independent interface for network control and management applications using agent software.

12.3.1 Block diagram

The design of the SwWA is depicted in the block diagram of Figure 12.3. All of its components are implemented as Java classes, apart from the CCM (Configuration Control and Management) library, which is a wrapper Java class for the implementation of low-level function calls specific to each switch model. These low-level function calls are implemented either in C or using the Java Management API and interact directly with the physical switch. The SwWA receives requests from a controlling agent (in the IMPACT architecture such agents are called Resource Agents or RAs (IMPACT, 1998a)) to perform a task on the switch. The methods do not block until the task is accomplished, but instead they store the task request in the queue of the Worker Thread corresponding to the calling RA and return immediately. Each Worker Thread corresponds to a different RA. The Worker Threads call methods of the *Virtual Switch* object which perform the requested task on the physical switch and also synchronise the state of the data structures of the Virtual Switch with the state of the physical switch.

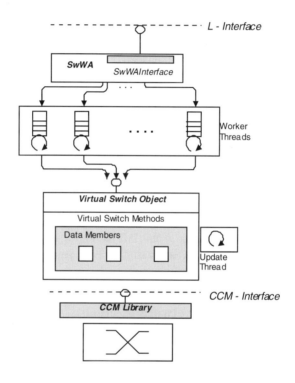

Figure 12.3   Block diagram of the SwWA

12.3.2   *The Virtual Switch Object*

The Virtual Switch class is a generic data model of an ATM switch's properties and physical and logical configuration. Only one Virtual Switch object is created per SwWA. It encapsulates appropriate data structures for representing configuration and performance information. These data structures can be handled only by the member methods of the Virtual Switch.

The member methods are called by every entity that wants to retrieve information from – or perform an action on – the physical switch. They first interact with the physical switch by calling methods of the CcmLib object (see below) and then perform the necessary updates (if needed) on the member data structures. All member methods constitute 'critical sections' of execution, i.e. only one thread at a time can enter any of the member methods, so as to preserve the consistency of the data structures with the physical switch state.

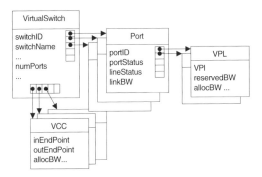

Figure 12.4   Member data structure of the Virtual Switch Class

The member methods can be divided into the following categories:

- *Bootstrap Synchronisation methods*: they are called on the initialisation of the Virtual Switch to synchronise the data structures with the current state of the physical switch.
- *Configuration and Connection Control methods*: they are called in order to perform a configuration management task or a connection control task.
- *Quantity Monitor Retrieval methods*: they are called to retrieve the current values of various operational parameters of the switch (stored in the Virtual Switch data structure).
- *Quantity Monitor Update methods*: they are called by the Update Thread in order to retrieve the current values of various operational parameters of the physical switch and update the data structures.

The SwWA has a switch-independent part (SI-part), between L- and CCM- interfaces, and a switch-specific part, which is the CcmLib. The SI-part has been implemented in Java and may be used for any switch type without modification. The CcmLib depends on the switch's proprietary MIB and has to be implemented for each different switch type. Currently, it has been implemented for Fore ASX switches and it is also being implemented for the Flextel switch.

12.4   Interaction with End-Systems Using ATM Signalling – the Proxy User Agent

The primary function of the Proxy User Agent (PUA) is to interface between Java RMI and ATM UNI signalling messages (ITU-T Recommendation Q.2931) and to take action depending on the contents of the messages. There is exactly one PUA for each terminal, i.e. a new PUA is invoked every time a new terminal is connected to an edge switch, and when a terminal is to be disconnected the corresponding PUA is killed.

12.4.1 The protocol layers underneath PUA

The protocol layers underneath the PUA are shown in Figure 12.5.

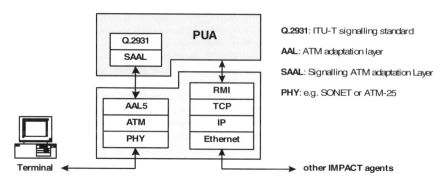

Figure 12.5 Protocol layers underneath the PUA

The left-hand protocol stack is used when communicating with the end terminal and the right-hand protocol stack is used when communicating with other IMPACT agents (e.g. peer PUAs). For simplicity, the inter-agent communication is carried over an Ethernet-based TCP/IP network.

12.4.2 The architecture of the PUA

The high-level architecture of the proxy user agent is shown in Figure 12.6.

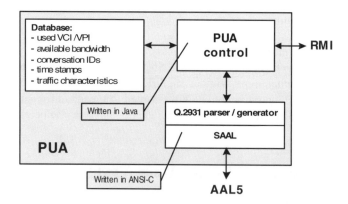

Figure 12.6 Architecture of the proxy user agent

The signalling processing part of the PUA has been implemented in ANSI-C whereas the high-level part of the PUA has been implemented in Java. The PUA contains a small database containing a list of the used VCI/VPI numbers at the UNI, all used connection identifiers, and a connection setup time stamp for each active connection. Furthermore, all relevant traffic characteristics are stored for each connection identifier.

The SAAL layer is based on ANSI-C code from the ATM-on-Linux project (1999), whereas the signalling part has been completely rewritten in order to include communication with the high-level part of the PUA.

12.4.3 Attaching the PUA to the terminal

Under normal circumstances, ATM UNI signalling is terminated in an ATM switch. In most cases, ATM switches do not allow ATM cells (or AAL5 PDUs) to be trapped or inserted at the network-side of the UNI. However, in order to be able to connect the PUAs to the terminals this is exactly what is needed. The solution to this problem is shown in Figure 12.7.

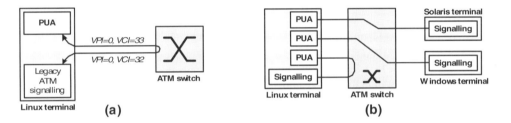

Figure 12.7   Attaching the PUA(s) to terminal(s)

The signalling software in the terminal is configured to use VCI=32 (and VPI=0) instead of VCI=5, which is normally used for signalling. Furthermore, the switch has a permanent loop-back connection (VCI=32 to VCI=33, and vice versa). In this way, the terminal has complete control of the ATM UNI signalling flow.

In the configuration shown in Figure 12.7(a), the terminal must be a Linux-based computer, because the low-level PUA parts are written specifically to run on Linux platforms. However, as shown in Figure 12.7(b), non-Linux terminals may also be used as long as the corresponding PUA is running on a Linux computer.

12.5 Running IP applications

Although IP over ATM was not in the main objectives of the project, IP applications are used for demonstration so as to show that the open control approach of IMPACT enables mixed IP and ATM environments with QoS support. One way of supporting IP over ATM is to use CLIP (Classical IP over ATM).

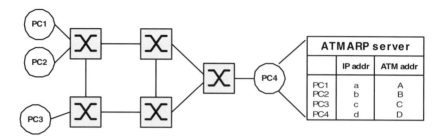

Figure 12.8   ATM network running CLIP

Consider the network shown in Figure 12.8. The terminal PC4 is set up to be an ATMARP[2] server. Each time a terminal (i.e. PC1, PC2, or PC3) is added to the CLIP network it starts a local ATMARP daemon, connects to PC4 (the ATMARP server), and registers itself. Thus, when all terminals have joined this CLIP network there is an ATM connection from PC4 to all terminals.

If, for instance, PC3 wants an IP (e.g. telnet or ftp) connection to PC1, it asks the ATMARP server for the ATM address of PC1. The ATMARP server sends back the requested ATM address, and finally PC3 sets up an ATM connection to PC1 using legacy signalling and the IMPACT agent system. However, all subsequent IP connections from PC3 to PC1 (or from PC1 to PC3) use the same ATM connection. This means that it is not possible to offer QoS to individual IP connections using CLIP.

12.5.1   Offering IP applications with Quality of Service (QoS)

To obtain on-demand ATM connections with specific traffic profiles and QoS through which IP application data will be sent, a *Terminal Agent* at the source IP terminal is invoked when the connection needs to be set up, just before an IP application is about to exchange data between source and destination.

The Terminal Agent (TermA) receives (by human user input) the traffic parameters and required QoS for the connection. It then sends a message requesting connection setup to the PUA at the access node. If the connection setup is successful, the TermA also configures the ATMARP daemon at the source terminal as is shown in Figure 12.9. A TermA also resides at the destination IP terminal and 'listens' to incoming ATM connections in order to configure the destination's ATMARP appropriately. After this procedure, the IP application can be initialised and the data exchange can begin. The TermA will interact with the PUA in two ways. Initially, it will use standard ATM signalling, being actually a signalling wrapper from the terminal's side. This capability is implemented using an API such as the Linux ATM sockets (ATM-on-Linux project, 1999). In a later stage, it will use the agent language provided by the Basic Agent Template, thus being able not only to convey

---

[2] ATMARP: ATM Address Resolution Protocol. The ATMARP server translates IP addresses into ATM addresses.

connection requests, but also to perform more 'intelligent' negotiations with the PUA either automatically or guided by a human user.

The example shown in Figure 12.9 indicates the Linux commands that enable ATMARP to be configured in the case of an X-Windows application running on workstation 10.0.0.7 with the DISPLAY exported to workstation 10.0.0.8. The application server (10.0.0.7 on the left) tells its ATMARP daemon to route all traffic destined for the IP number 10.0.0.8 through an ATM VC with VPI=0 and VCI=77. At the same time, it configures the network interface called 'atm8' to have IP address 10.0.0.7. In the same manner, the client terminal configures a network interface called 'atm5' to have the IP number 10.0.0.8, and a direct connection to 10.0.0.7 can be obtained through VPI=0 and VCI=77. The ATM connection between the two hosts could be set up using the IMPACT agent system.

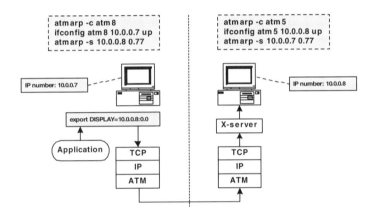

Figure 12.9   Setting up point-to-point X-Windows applications over ATM

12.6 Conclusions

In order to implement an agent system for network resource control, an open framework for the deployment of network control schemes has to be provisioned first. This framework will enable the agents to access the low-level capabilities of network devices, bypassing the control logic embedded by the manufacturers. This chapter has presented the implementation of such a framework in an ATM network environment. However, the concepts introduced are fairly generic and can be applied to IP networks. This applicability is almost direct, when considering enhancements to the IP service model that offer QoS (such as those described within the IntServ and DiffServ frameworks). A major concern regarding deploying this framework and, furthermore, the agent-based control system, relates to the control performance, since management protocols are used instead of low-level control software in case the network devices do not offer some means of programmability. However, after having conducted several experiments with the system, it is our belief that the potential benefits in terms of flexibility and more efficient resource usage can compensate for some performance degradation.

12.7   References

ATM-on-Linux Project. (1999) Web site: http://lrcwww.epfl.ch/linux-atm [13 June 2000].

Biswas, J. et al. (1998) The IEEE P1520 Standards Initiative for Programmable Network Interfaces. *IEEE Communications Magazine*, October, 64–71.

IMPACT (1998a) Specification of Agents. IMPACT AC324 Project Deliverable 03. Web site: http://www.acts-impact.org [13 June 2000].

IMPACT (1998b) *Specification of Signalling Implementation.* IMPACT AC324 Project Deliverable 04. Web site: http://www.acts-impact.org [13 June 2000].

Laubach, M. (1994) Classical IP over ATM. RFC 1577, IETF, Standards Track, January 1994.

Soldatos, J., Kormentzas, G., Vayias, E., Kontovasilis, K. and Mitrou, N. (1999) An Intelligent Agents-Based Prototype Implementation of an Open Platform Supporting Portable Deployment of Traffic Control Algorithms in ATM Networks, in *Proceedings of the 7th COMCON Conference*, Athens, Greece.

Vayias, E., Soldatos, J., Kormentzas, G., Mitrou, N. and Kontovassilis, K. (1998) Monitoring Networks over the Web: Classification of Approaches and an Implementation, in *Proceedings of the 4th International Conference on Telecommunications,* pp. 451–456, Chalkidiki, Greece.

13

A Multi-Agent Approach to Dynamic Virtual Path Management in ATM Networks

P.Vilà, J. L. Marzo, R. Fabregat, D. Harle

13.1   Introduction

Most currently available management tools and frameworks are based around a centralised approach and do not take advantage of the technology they are designed to manage. Usually, human operators manage these networks at a variety of organisational levels. Current management platforms are confronted with a scalability problem as user needs grow.

This chapter describes a multi-agent system designed for Virtual Path (VP) capacity management and restoration. Since this is quite a specific application, in section 13.2 we explain a few important aspects of communications theory, focusing on ATM networks and Virtual Path management, to illustrate why these problems may be suitably addressed using a multi-agent system; the context of our multi-agent system is also introduced. In section 13.3, we introduce the problem specification, in particular VP capacity management and restoration. In section 13.4, we present some previous approaches concerning agents and ATM management. In section 13.5, the multi-agent system itself is presented, and in section 13.6, we propose an experimental testbed. Finally, in section 13.7, we draw some conclusions about this proposed solution.

13.2   Network Management

Network management is a huge field comprising: fault, accounting, configuration, performance and security management. These different facets of network management are closely interrelated; for example, if a fault occurs, it has an effect on both performance and the network configuration, until the system isolates the fault and continues its normal operation. In general, current network management systems are centralised.

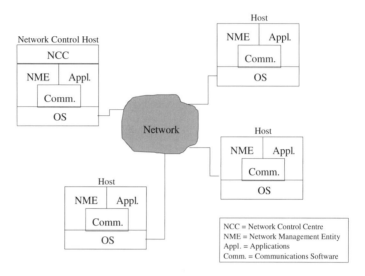

Figure 13.1   A typical network management system architecture

Figure 13.1 shows the architecture of a typical network management system. Each network node contains a collection of software devoted to the task of network management, which is referred as the Network Management Entity (NME) in the figure. Each NME performs the following tasks:

- Collects statistics on communications and network-related activities
- Stores statistics locally
- Responds to commands from the network control centre, including commands to transmit collected statistics to the network control centre, change a parameter, provide status information, and generate artificial traffic to perform a test.

At least one host in the network is designed as the Network Control Centre (NCC). NCC-NME communication uses an application-level network management protocol that employs communications architecture in the same way as any other distributed application.

Well-known management standards (Black, 1994) include the International Telecommun-ications Union – Telecommunications (ITU-T), recommendations of the Telecommunications Management Network (TMN) (Johnson, 1993) and the Internet management standard Simple Network Management Protocol (SNMP) (Stallings, 1996; Stallings, 1998). All are centralised approaches whose main limitations are their scalability as the network grows and the bottlenecks that arise from a centralised vision of the network.

13.2.1 Overview of ATM

Asynchronous Transfer Mode (ATM) has been proposed as the underlying transport and switching technique for the Broadband Integrated Services Digital Network (B-ISDN). As such, these networks are designed to support a variety of services with diverse characteristics, e.g. voice, video and data traffic, etc. ATM is a connection-oriented packet-switched data transport system, handling short fixed-size packets called cells. These fixed-size cells are asynchronously multiplexed within the network and transmitted over a virtual circuit while preserving their cell sequence integrity. ATM offers a complete and flexible network solution whose main advantages are:

- ATM is a scalable technology. The ITU-T standards define the format of the 53-byte cell, but they do not define items such as rates, framing or physical bearers. Many different systems, e.g. LANs, WANs, and public networks, use the same cell format which eases interconnection between networks.
- ATM is service-transparent. It can accommodate a wide range of services with diverse characteristics, e.g. voice, data and video traffic, and can provide bandwidth at virtually any rate above 64 kbit/s.
- ATM is bandwidth-efficient. Several connections are multiplexed onto the same physical link, and ATM uses statistical multiplexing to take advantage of the large peak-to-average bandwidth ratios present in some services.

13.2.2 Traffic and congestion control

Traffic control and resource management have the crucial role of protecting the network from becoming congested, achieving the network performance objectives and optimising the use of network resources. Network provisioning (NP) is a set of long-term control actions that determine the physical quantities of resources (buffers, links, etc.) to be placed in the network. The challenge is to ensure that sufficient resources are available to accept all potential connections.

Traffic management functions can be divided into two broad classes: traffic control and congestion control functions. Traffic control refers to the set of actions taken by the network to avoid congested situations, and congestion control refers to the set of actions taken by the network to minimise the intensity, spread and duration of congestion. In other words, traffic control techniques are preventive, and congestion control techniques are reactive.

ATM network management requires both efficient preventive and efficient reactive techniques. Reactive techniques alone are not sufficient, especially for WANs and public networks, where the large propagation delay/bandwidth product means that feedback algorithms (i.e. reactive mechanisms) are ineffective.

A number of traffic and congestion control mechanisms have been proposed:

- *Network Resource Management* (NRM). The network may be dimensioned such that resources are allocated in order to separate traffic streams according to their service characteristics. A network resource is any component of the network, hardware or software, which supports the communication needs. The two generic resources that are spread throughout the network are time resources (link rate or bandwidth) and space resources (buffer space). NRM is concerned with short-or-mid term resource configuration/allocation.

- *Connection Admission Control* (CAC) is responsible for determining whether a connection request should be accepted or rejected by the network. Connection Admission Control can be defined as 'the set of actions taken by the network during the call setup phase (or during the call renegotiation phase) to establish whether a virtual channel connection or a virtual path connection can be accepted or rejected' (McDysan and Spohn, 1994).

- *Usage and Network Parameter Control* (UPC/NPC). This function ensures that the traffic (on each connection) remains within the parameters negotiated at the connection setup phase. UPC is performed at the user network interface whereas the NPC is performed at the network-node interface. UPC can take different actions if a source violates its contract. It can delay violating cells or simply drop them, it can change their CLP bit (in the header of the cell) so that they can be dropped, if required by the intermediate nodes along the virtual channel, or UPC can inform the source when it starts to violate its contract.

- *Priority Control*. A source can generate traffic flows with different priorities by using the CLP bit. A congested network element can selectively discard low-priority cells if necessary to protect high-priority cells.

- *Traffic Shaping*. Traffic shaping is performed at the user network interface it manipulates the flow of data so that it respects the negotiated contract. In other words, traffic sources use traffic shaping to process the cell stream and ensure that the resultant output satisfies the contract negotiated at the connection setup phase. Traffic shaping can be used to reduce the peak cell rate, limit burst length, or reduce cell delay variation by suitably spacing cells in time.

- *Fast Resource Management*. Fast Resource Management responds to a user request to send a burst and allocate capacity (e.g. bandwidth, buffer space) for the duration of the burst along the virtual channel. There are two possible modes for bandwidth reservation: immediate unguaranteed and delayed guaranteed.

- *Selective Cell Discarding*. In order to protect high-priority, non-violating cells, a congested network element may selectively discard cells which are explicitly identified as belonging to a non-compliant ATM connection and/or cells which have their CLP bit set to one, i.e. low-priority cells.

- *Explicit Forward Congestion Indication*. A congested network element may set the Explicit Forward Congestion Indication bit in the header of the cells. The receiving node can then use this information and lower the cell rate of the sending node.

In time scales, NP is the long-term physical resource allocation, while NRM is a short-or mid-term resource configuration or allocation. NRM is applied on a service level or on groups of connections, e.g. VPs, while CAC is based on individual connections. Routing operates either as part of or in parallel with all three control levels. The relationships between traffic control functions are schematically shown in Figure 13.2.

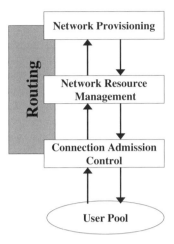

Figure 13.2   Relationships between ATM traffic control functions

In order to guarantee the proper operation of traffic control and congestion control functions, the intrinsic characteristics of connections need to be adequately described by a set of standardised traffic parameters.

13.2.3   ATM connections

ATM connections that have homogeneous characteristics in terms of traffic patterns, quality of service (QoS) requirements, and possible use of control mechanisms, are grouped in several different classes of service. Broadly speaking, these service classes may be classified into the following traffic classes:

- Guaranteed services, Constant Bit Rate (CBR) and Variable Bit Rate (VBR) with fixed traffic descriptors, whereby the network provides a QoS guarantee as long as the traffic source respects its negotiated parameters.
- Available Bit Rate (ABR), where the source provides minimum traffic requirements and dynamically adapts to feedback from the congestion control mechanism.
- Unspecified Bit Rate (UBR) where no traffic descriptors are used and no QoS guarantees are made.

ATM networks have several layers of hierarchy. In the ATM layer there are two levels of hierarchy: Virtual Path (VP) and Virtual Channel (VC) levels. The highest is the VC level – users can establish and release connections, i.e. Virtual Channels, through pre-established VPs. The ATM cells flow through the VCs, as is shown in Figure 13.3.

A VP is essentially a pair of VPCs used to achieve bidirectional communication, each using the same VP identifier (VPI) in both directions. A VPC is a concatenation of VP links carrying a unidirectional flow of cells belonging to a VP. A VP link is a unidirectional flow of cells between two consecutive terminating points, i.e., ATM switches and cross-connects.

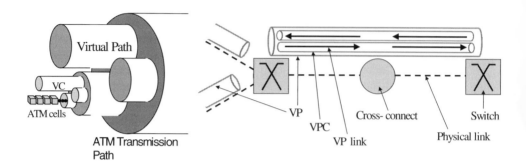

Figure 13.3   Hierarchical layers of an ATM network

All VPs require a capacity assignment and are used in order to simplify both the establishment of new VCs and statistical multiplexing; they are also used to constitute a Virtual Path network, i.e. a virtual topology over the physical one (see Figure 13.4). The main characteristic of this logical network is its flexibility, since it is independent of the physical network; this VP network and VP capacities can be reassigned dynamically so as to maintain the VP network with certain desirable characteristics.

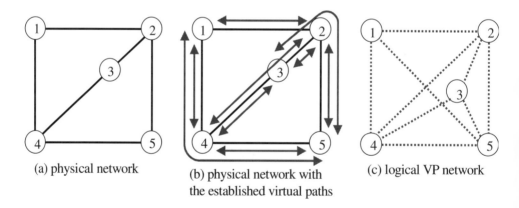

Figure 13.4   Physical and logical network topologies

The VP management considered is the establishment and release of VPs to accommodate traffic demands, dynamic bandwidth management to optimise network utilisation, and restoration when a fault occurs. Many of these operations are semi-permanent and are normally performed by a centralised entity.

13.3   Problem specification

13.3.1   VP capacity management

Bandwidth management attempts to manage the capacities assigned to the different VPs that flow through a physical link. Sometimes, parts of the network become under-utilised and other parts congested. When this occurs, some connections are rejected that could be accepted if the traffic load were better balanced. The network rejects a connection when there is insufficient free capacity in the VP to be traversed, because existing connections are employing most of the bandwidth allocated to this VP.

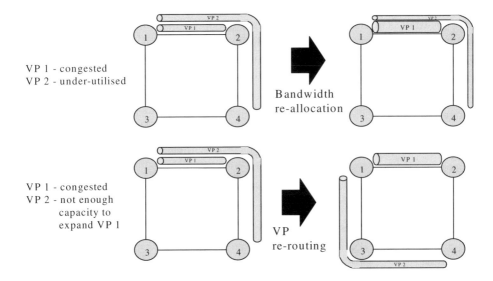

Figure 13.5   Possible actions to avoid congestion

Figure 13.5 illustrates the two actions normally taken by the bandwidth management system:

*   If there are congested VPs and underused VPs in the same link, the bandwidth assigned to each VP can be reconfigured so that the worst call-blocking probability in any VP is minimised. This method is called bandwidth reallocation.
*   If almost all the VPs in the link are congested or near congestion and there is insufficient unutilised bandwidth capacity for swapping between VPs, routes as well as capacities are altered to maximise the traffic carried in the network. This means that a change in VP network topology is required. This is called VP re-routing. In some cases, when a topology change is required, re-routing is used to calculate the overall optimal redistribution of the VPs in order to cope with the traffic demands. In other cases, the

priority is to minimise the number of re-routed VPs so as to minimise the number of connections affected.

These operations are normally performed, at a general network level, by a central operations system. The system collects traffic-load and VP (bandwidth and routes) information for the entire network, applies an algorithm to this information and generates the appropriate control actions. The new routing tables and/or new bandwidth allocations are then downloaded to the appropriate nodes within the network.

Whether caused by inaccurate forecasting or unexpected overloads, individual VPs become under- or over-utilised at particular instances. Centralised schemes react by using the above mechanisms at periodic control intervals. The timing interval, however, is crucial: activating these mechanisms over longer intervals (of the order of hours) may result in the network management system missing and not responding to instances of capacity shortfall. On the other hand, when adopting short intervals, the value of centralised bandwidth management is questionable due to the processing overheads. To overcome this dilemma, we propose to apply these schemes in a dynamic responsive manner, i.e. when the system detects the problem or the conditions that will shortly cause a problem.

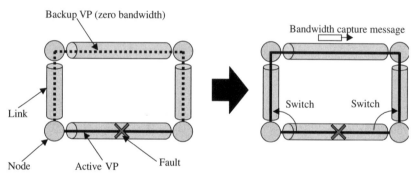

Figure 13.6   Switching from a failed active VP to the backup VP

Pre-planned schemes are based on pre-assigned backup VPs, whereas dynamic schemes are based on flooding algorithms and search for restoration routes by broadcasting messages after a failure is detected. Pre-assigned backup VPs can have zero bandwidth when established because of the independence between route establishment and bandwidth setting in ATM networks. Spare resources offered by a link are shared between backup VPs accommodated by that link and managed by the node connected to the link. If a network failure occurs, bandwidth is reserved using distributed message communications between nodes on the backup VP which follows the detection of that failure. After the appropriate bandwidth is reserved in all links on the backup VP, the failed active VP is switched to the backup VP (Figure 13.6).

Hybrid mechanisms can also be used, i.e. both dynamic and pre-planned schemes. Yahara and Kawamura (1997) propose different levels of protection depending on the service quality or protection level that the user wants or can afford (Figure 13.7). This scheme can be understood as a priority assignation among the VPs.

All these functions (VP establishment and release, VP dynamic bandwidth management and VP hybrid restoration, as well as the spare bandwidth to guarantee some level of restoration) have to be managed. Their mutual interference (e.g. a fault causing re-routing of some VPs, and this causing both bandwidth and spare capacity adjustment) means that the management of VPs in ATM networks becomes extremely complex. In the following section, we propose a multi-agent system to implement these functions in ATM networks.

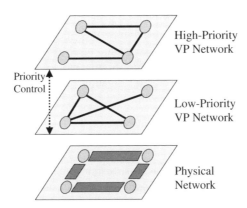

Figure 13.7   Multi-reliability VP network

13.4   Intelligent Agents

13.4.1   Intelligent software agent characteristics

An *intelligent agent* or *intelligent agent system* is a software-based computer system that enjoys the following properties:

- Autonomy: agents operate without the direct intervention of humans or others, and have some kind of control over their actions and internal state.
- Social ability: agents interact or cooperate with other agents (and possibly humans) via some kind of agent communication language.
- Reactivity: agents perceive their environment, and respond in a timely fashion to changes that occur in it.
- Pro-activeness: agents do not simply act in response to their environment, but are able to exhibit goal-directed behaviour by taking the initiative.

Other attributes that can be considered in the intelligent agent systems are mobility (the ability of an agent to move around a network), veracity (the assumption that an agent will not knowingly communicate false information), benevolence (the assumption that agents do not have conflicting goals and that every agent will therefore always try to do what is asked of it), and rationality (the assumption that an agent will act in order to achieve its goals).

From the AI perspective, an intelligent agent system is thought of as a computer system that, in addition to having the properties identified above, is either conceptualised or implemented using concepts that are more usually applied to humans, i.e., mentalistic notions, such as knowledge, belief, desire, intention and obligation. It is worth considering exactly which attitudes are appropriate for representing agents. Information attitudes (knowledge, belief) are related to the information that an agent has about the world it occupies, whereas pro-attitudes (desire, intention, obligation, etc.) are those that in some way guide the agent's actions. It seems reasonable to suggest that an agent must be represented in terms of at least one information attitude, and at least one pro-attitude (Nwana, 1996; O'Brian and Nicol, 1998).

13.4.2   Agent architectures

Agent architectures can be defined as a particular methodology for building agents. They specify how the agent can be decomposed into the construction of a set of components or modules and how these modules should be made to interact. The total set of modules and their interactions has to provide an answer to the question of how the sensor data and the current internal state of the agent determine the actions and future internal state of the agent.

A classical approach is to adopt a deliberative architecture. A deliberative agent or agent architecture is defined as one that contains an explicitly represented, symbolic model of the world, and in which decisions (for example, about what actions to perform) are made via logical (or at least pseudo-logical) reasoning, based on pattern matching and symbolic manipulation.

Reactive architectures represent an alternative to that offered by the classical approach. A reactive architecture is defined as one that does not include any kind of central symbolic world model, and does not use complex symbolic reasoning. These architectures are based around three ideas: that intelligent behaviour can be generated without explicit representations of the kind that symbolic AI proposes; that it can be generated without explicit abstract reasoning of the kind also proposed by symbolic AI; and that intelligence is an emergent property of certain complex systems. It can be said that such architectures act using a stimulus/response type of behaviour by responding to the present state of the environment in which they are embedded

A hybrid architecture is one that combines both classical and alternative approaches. The approach proposed is to build an agent framework consisting of two (or more) subsystems based around deliberative and reactive architectures. A deliberative system contains a symbolic world model, which develops plans and makes decisions in the way proposed by mainstream symbolic AI. A reactive system is capable of reacting to events that occur in the environment without engaging in complex reasoning. Often, the reactive component is given some kind of precedence over the deliberative one, so that it can provide a rapid response to important environmental events. This kind of structuring leads naturally to the idea of a layered architecture.

13.4.3   Agents and ATM management

There has been significant research into intelligent agents and it can be seen that network management and ATM management, in particular, are ideal candidates to employ such mechanisms. There are many examples of intelligent agents for network management and a brief introduction to relevant work is given. Further information can be found in Hayzelden and Bigham (1999).

In the HYBRID project (Somers, 1996; Somers, Evans, Kerr and O'Sullivan, 1997), a well-defined architecture based in a geographical hierarchical structure in three layers – local, regional and national – is described. Each of the layers is responsible for a particular region of the network. The upper layers delegate the management to the lower ones and when there is a problem the lower layers cannot solve it is sent to the upper layers. The main difference of the approach proposed in this chapter is that the proposed architecture is more oriented to VP management whilst the agents are dynamic and distributed in a logical manner.

In Lucent Technologies (1997), a more general situation is described which focuses more on learning and predicting agents based on multi-variate statistical analysis techniques and the design of experiments.

The work of Davison, Hardwicke and Cox (1998) focuses solely on VP dynamic bandwidth management (it does not carry out any kind of fault management, for example) and uses two completely independent multi-agent systems working in parallel. There is no global network view or planning, stressing the simplicity of the agents.

An approach based on multiple layers of multi-agent systems is proposed in the Tele-MACS architecture (Hayzelden and Bigham, 1998; Hayzelden, 1998), which integrates distributed low-level reactive agents and pro-active planning agents. The agent system subsumes the proprietary control system with the purpose of providing a greater degree of intelligent control. The main difference of the proposed approach is that it utilises different classes of reactive agents, each class controlling a single management function of the network.

The ACTS (Advanced Communication Technologies and Services) project IMPACT (1998) presents a more complex scenario where a multi-layered agent system is designed to carry out complete ATM resource management including Call Admission Control (CAC), routing, dynamic bandwidth management, QoS management, etc. The idea is demonstrate that agent-based control can act within the constraints of traditional connection admission procedures in a real ATM testbed.

13.5   Multi-Agent System Proposal

An ATM VP management system based on intelligent software agents is proposed. The main characteristics of this system are: scalability, robustness and simplicity. The main goals are to maximise the autonomy of the agents and to minimise the communications between them. The architecture of this system is based on two different independent multi-agent systems (MAS). They are denoted as the Network Monitoring MAS (NM-MAS) and the Network Planning MAS (NP-MAS).

Common characteristics of these types of MAS are the above-mentioned robustness and scalability. Robustness implies a certain duplication of resources, i.e. more than one agent monitors a VP, watching, for example, for the possibility of node failure. In this case, scalability also means the reduction of the network traffic, i.e. the communications between agents. Apart from minimising, negotiation is only carried out when necessary and transferring knowledge instead of bulk data. An interesting restriction applied in Lucent Technologies (1997) is that the agents situated in the nodes, where the computing facilities are, can only communicate with other agents in the same node and with their neighbours. These neighbours are, of course, virtual neighbours. Two agents, at both terminations of a VP or physical link, are neighbours. This facilitates the goal which is to develop simple distributed techniques that scale well for very large networks. In Figure 13.8, agent A1 in node 3 can communicate with agent A2 in node 2, but not with any agent in node 4.

The same figure also illustrates the idea of replication to achieve robustness. Even if the links between the nodes are physical links or VPs, two agents control every link or VP.

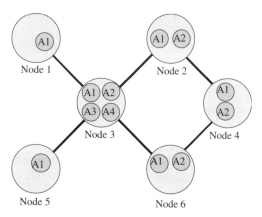

Figure 13.8   Agent neighbourhoods

13.5.1   Network Monitoring – Multi-Agent System (NM-MAS) architecture

The functions of NM-MAS agents are to monitor and control the network. The main goal is that they are simple and react quickly when an event (connection attempt, connection release, load change, fault, etc.) occurs. To achieve these goals, a purely reactive structure is proposed. Intelligence is an emergent property in this kind of system with a large number of small distributed agents interacting with each other (Wooldridge and Jennings, 1995). These agents act by using a stimulus/response type of behaviour to respond to the state of the environment in which they are embedded. Therefore, a simple architecture based on rules is proposed for this kind of agent.

As the goal is to implement VP networks with different levels of priority, the NM-MAS agents are also grouped at different levels of priority. An example of the proposed architecture is shown in Figure 13.9. Here it can be seen that there are 8 agents in the physical network monitoring and controlling the physical links. Above the physical layer in the low-priority VP network, there are 8 other agents. In the high-priority network, there are also 8 other agents, each monitoring and controlling the high-priority VPs. Agents are located within the nodes of each of the differing-priority VP (logical) networks and the physical network, but in fact there is only one physical node. Thus, the agents in the different VP networks reside physically within the same node. For example, in node 1 there are two agents in the physical network, three in the low-level VP network, and one at the high-priority VP level. In real ATM networks there will be a large number of VP networks, links and priority levels, resulting in a large number of NM-MAS agents; these agents must be simple.

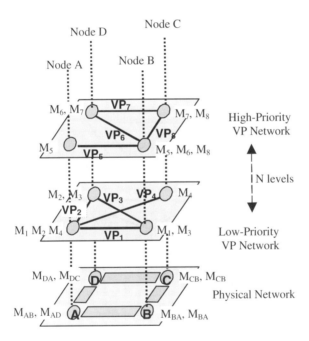

Figure 13.9   Network Monitoring – Multi-Agent System (NM-MAS) architecture

13.5.2   *Network Planning – MultiAgent System (NP-MAS) architecture*

The task of these agents is to monitor and control the whole network through monitoring the NM-MAS agents; they do not monitor the network directly. These agents consult the NM-MAS agents' knowledge and modify their rules and goals by including the NP-MAS information. Another task of this system is to maintain some distributed overall view of the network through the communication of each agent with its neighbours (the scalability restriction). There is only one NP-MAS agent in each node. These agents must be bigger and more complex than NM agents and must also have reasoning and planning subsystems.

This means that a different architecture is required for the NP-MAS agents. As their goal is to maintain a distributed overall view of the network in order to make plans and forecasts, a collaborative approach is adopted. This means that each agent operates over a limited domain and has limited information and resources but, by pooling together their abilities, agents are able to solve problems beyond the capacity of any one single agent. The proposed architecture is a good example of a layered architecture (Collins, Ndumu, Nwana and Lee, 1998).

As the proposed agent systems are not yet required to communicate with other different agent systems, standard agent communication protocols (Nwana and Wooldridge, 1996), e.g. Knowledge Query and Manipulation Language (KQML) (Somers, 1996; Somers et al., 1997) are not yet to be deployed. This decision also makes it easier to implement the system initially. However, as in O'Brian and Nicol (1998), KQML is criticised and another Agent Communication Language (ACL) is proposed the choice of a standard ACL can be postponed until a later date.

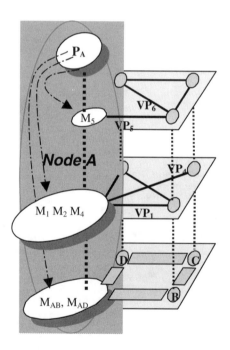

Figure 13.10   Network Planning – Multi-Agent System (NP-MAS) architecture

13.6   Experimental Testbed

The throughput of a network supported by the proposed multi-agent system when subject to different loads is to be investigated. A software tool to emulate an ATM node underpins the experimental setup. Although it is intended to define the test scenarios to be as realistic as possible, there is no direct need to utilise a real ATM node. The main interest is in a subset of the functions performed by the node, and software emulation is deemed sufficient.

Each node supports three different systems, as can be seen in Figure 13.11. The Traffic Event Generator (TEG) simulates the user's behaviour; each TEG system is able to simulate new connections to be established, to be released, and to change the instantaneous load dynamically, etc. TEG simulates ON OFF traffic; the activity can be configured by setting up some parameters such as interarrival time and the sojourn time for each type of connection. Traffic parameters can be also configured, typically in terms of mean and peak rate and mean burst rate, and these parameters allow the utilisation of bandwidth allocation strategies to be investigated. The system can then be used to evaluate the network management behaviour under different load conditions.

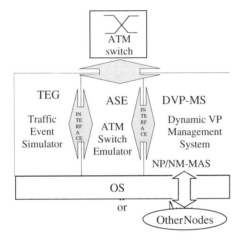

Figure 13.11   The ATM Node Model supporting its three different subsystems

The ATM Node Model emulates the operation of an ATM switch. This system contains all the information on the VP and VC configurations of its node and receives events and notifications from the TEG.

The Dynamic VP Management System (DVPMS) adapts the resources on the network based on the node configuration information and the inter-agent coordination. This system is made up of the two multi-agent subsystems described previously (NM-MAS and NP-MAS).

There are two separate levels of communication between nodes through a distributed platform, one level for the Node Model communications (classical ATM control – CAC and Routing) and another one for the multi-agent system communication.

In order to carry out experiments the following scenarios are defined: a) VPs are previously established, the initial BW allocated may be dynamically changed; b) users communicate with traffic control (CAC and Routing) as in classical environments; and c)

agents communicate with traffic control and with the other agents in order to reallocate resources. They cannot directly communicate with users.

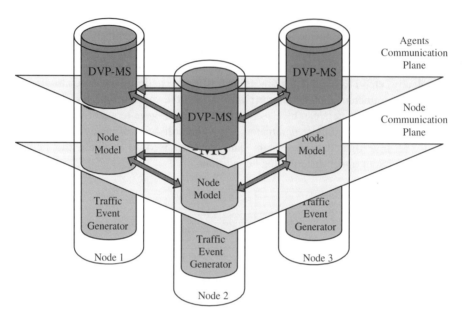

Figure 13.12   The experimental testbed

The key functionalities of each node element are summarised in Table 13.1 below.

Table 13.1   Functions of node elements

Functions	Origin	Description
Events	TEG	Traffic Event Generator (TEG) simulates new connections set up, releases, and, eventually, loads variations. Events are passed to the Node Model as independent messages.
Peer commands	Node Model	Node-to-Node commands asking for routes, current load, or other information at DVP-MS level.
Human commands	Human	Used to configure and monitor the network by human manager. The console is used to interact with DVP-MS.
Set up commands	NP and Human	Agents use setup commands to allocate resources dynamically. (All of these commands can be invoked as human commands for debugging.)

13.7   Conclusions and Future Work

In this chapter, a specific aspect of ATM network management has been discussed, namely VP capacity management. Ideally suited to the use of distributed artificial intelligence techniques and in particular software agents, a framework based upon these techniques has been proposed. The proposed framework will enable a distributed, reactive and flexible approach to be applied to the specific tasks of VP capacity management and VP restoration. Additionally, an experimental testbed to enable the evaluation of this framework has also been proposed.

The completion of the platform implementation represents the next stage of work, followed by an in-depth performance evaluation to verify its operation and effectiveness. In the first instance, a fully functioning NM-MAS along with a basic-functionality NP-MAS system will be used. Such a basic system will then allow the introduction of new modules into the NP-MAS to improve its performance and increase its repertoire. For example, experiments with a more powerful forecasting module for better network planning can be done.

Another area of interest is the human interface to the management system, i.e. how the human network manager interacts with the system in order to introduce his/her instructions into the network. Network management in general, and the ATM VP management, in particular, is increasing in complexity in such a rate that it is becoming more and more difficult for the human operator. An important goal is to construct network management with no human intervention for day-by-day management, with only special cases left to the human manager. This means that the network manager's instructions and policies still have to be translated into a consistent set of goals and knowledge for each agent, and this is still to be addressed.

13.8   References

Bigham, J., Cuthbert, L.G., Hayzelden, A.L.G. and Luo, Z. (1999) Multi-Agent Systems for Resource Management, in H. Zuidweg, M. Campolargo, J. Delgado and A.P. Mullery (Eds.) *Intelligence in Services and Networks – Paving the Way for an Open Service Market (*Lecture Notes in Computer Science 1597*)*, pp. 514–526, Springer, ISBN 3-540-65895-5.

Black, U. (1994) *Network Management Standards.* McGraw-Hill, ISBN 0-07-005570-X.

Collins, J.C., Ndumu, D.T., Nwana, H.S. and Lee, L.C. (1998) The ZEUS agent building tool-kit. *BT Technology Journal*, **16(3)**.

Davison, R.G., Hardwicke, J.J. and Cox, M.D.J. (1998) Applying the agent paradigm to network management. *BT Technology Journal*, **16(3)**.

Hayzelden, A.L.G. (1998) Telecommunications Multi-Agent Control System (Tele-MACS), in *Proceedings of the 13th European Conference on Artificial Intelligence (ECAI 98)*.

Hayzelden, A.L.G and Bigham, J. (1998) Heterogeneous Multi-Agent Architecture for ATM Virtual Path Network Resource Configuration, in S. Albayrak and F. Garijo (Eds.) *Intelligent Agents for Telecommunications Applications,* Springer, ISBN 3-540-64720-1.

Hayzelden, A.L.G. and Bigham, J. (1999) Agent Technology in Communications Systems: An Overview. *The Knowledge Engineering Review*, **14(3)**, 1–35.

IMPACT (1998) ACTS Project AC324. Web site: http://www.acts-impact.org [13 June 2000].

Johnson, P.A. (1993) Domestic and International Open Systems Interconnection Management Standards, in S. Aidarous and T. Plevyak (Eds.) *Telecommunications Network Management into the 21st century*, IEEE Press, ISBN 0-7803-1013-6.

Lucent Technologies. (1997) Proactive Problem Avoidance and QoS Guarantees for Large Heterogeneous Networks. © Lucent Technologies, Rensselaer Polytechnic Institute and Pennsylvania State University.

McDysan, D.E. and Spohn, D.L. (1994) *ATM Theory and Application*. McGraw-Hill.

Nwana, H.S. (1996) Software Agents: An Overview. *The Knowledge Engineering Review*, **11**.

Nwana, H.S. and Wooldridge, M. (1996) Software Agent Technologies. *BT Technology Journal*, **14(4)**.

O'Brian, P.D. and Nicol, R.C. (1998) FIPA – towards a Standard for Software Agents. *BT Technology Journal*, **16(3)**, 51–59.

Sanz, R. (1998) Methodologies for Complex Control Systems Engineering, in *Proceedings of the COSY Workshop on Integration of Complex Systems, Zurich*.

Somers, F. (1996) HYBRID: Unifying Centralised and Distributed Network Management Using Intelligent Agents, in *Proceedings of the IEEE/IFIP Network Operations and Management Symposium (NOMS'96)*.

Somers, F., Evans, R., Kerr, D. and O'Sullivan, D. (1997) Scalable Low-Latency Network Management Using Intelligent Agents, in *Proceedings of the ISS'97, XVI World Telecom Congress*.

Stallings, W. (1996) *SNMP, SNMPv2 and RMON: Practical Network Management*. Addison-Wesley, ISBN 0-201-63479-1.

Stallings, W. (1998) SNMP and SNMPv2: The Infrastructure for Network Management. *IEEE Communications Magazine*, March.

Wooldridge, M. and Jennings, N.R. (1995) Intelligent Agents: Theory and Practice. *The Knowledge Engineering Review*, **10(2)**, 115–152.

Yahara, T. and Kawamura, R. (1997) Virtual Path Self-Healing Scheme Based on Multi-Reliability ATM Network Concept, in *Proceedings of IEEE GLOBECOM'97*.

14

A Multi-Agent Approach for Channel Allocation in Cellular Networks

E. L. Bodanese, L. G. Cuthbert

14.1   Introduction

Previous work on analogue and second-generation mobile communications has led to several schemes being proposed to maximise the channel usage and minimise the call blocking. Some channel assignment schemes presented in the literature (e.g., Elnoubi, Singh and Gupta (1982), Zhang and Yum (1989), Karlsson and Eklundh (1999), Jiang and Rappaport (1994), Das, Sen and Jayaram (1997), Das et al. (1997) and Katzela and Naghshineh (1996)) have improved the performance of the basic fixed channel assignment strategy for different traffic densities (macro-micro-pico-cellular networks) over different traffic load conditions. However, most of the solutions proposed have an entirely reactive approach – the response to a series of events follows an algorithm that is prepared to react to specific situations – which limits their efficiency. Even those schemes that contain adaptive features are not ideal. Some are completely centralised with high computational complexity and signalling overhead, making the implementation in real systems barely possible, while distributed adaptive schemes are restricted to individual base stations without cooperation features.

In order to accommodate multiple networks and services with multiple bit rates within a limited frequency band, resource flexibility is one of the most important requirements in third-generation networks. A multi-agent system is able to provide greater flexibility and also increase the robustness of the network by allowing negotiation when conflicts occur. This chapter describes a framework we are adopting to verify the feasibility of multi-agent systems in controlling resource assignment in mobile networks.

We are assuming a macrocellular scenario where base stations are not able to share information by interference measurements, but only by explicit exchange of information, and the resources are complete frequency carriers. Although the results in this chapter are based on an AMPS model (MacDonald, 1979), the concept is completely generic and the work could be extended to TDMA/CDMA. Different cell structures (such as microcellular or hierarchical) could also be included.

14.2   A Reactive Channel Allocation Scheme Implementation and Analysis

For macrocellular systems, where explicit communication is needed, FCA with channel borrowing offers good results and less computational complexity than DCA. However, those FCA schemes with the best results (e.g. BCO (Elnoubi et al., 1982) and BDCL (Zhang and Yum, 1989)) use centralised control inside the Mobile Switching Centre (MSC), where there is still a need to maintain an up-to-date global knowledge of the entire mobile network. To alleviate this problem, several authors have proposed modifications to make the schemes more distributed. One example is D-LBSB (Das et al., 1997) which performs better than its centralised version (Das, Sen and Jayaram, 1997) and also outperforms other existing schemes like direct retry (Karlsson and Eklundh, 1999) and CBWL (Jiang and Rappaport, 1994). D-LBSB is a distributed FCA algorithm with selective borrowing, channel locking and channel reassignment. It takes into consideration the position of the mobile users when borrowing and reassigning channels; it triggers the execution of the algorithm when the usage of the nominal channels in a cell reaches a pre-determined threshold (h, when a previously *cold* cell becomes *hot*). It also controls the number of the channels to be borrowed from or lent to a cell, according to the traffic load of the whole cellular network. We selected D-LBSB (Das et al., 1997) as the basis for a scheme implementation, because it follows a complete reactive algorithm and hence provides us with an example to be analysed. It also serves as a basis for comparison with the multi-agent system over the same simulation context and scenarios.

We have implemented the cellular model in the commercial simulator OPNET™. There are 49 cells and each cell has 10 nominal channels. The compact pattern is a 7-cell cluster with the reuse distance being three cell units. Mobile users have their own trajectories inside the mobile network. Call establishments and handoff requests are simulated as they are requested in AMPS systems. For each cell, a Poisson distribution is used to generate calls, which have an exponentially distributed duration with a mean of three minutes. An idle mobile inside the cell performs the call attempt. This is a more detailed simulation than those presented in the literature, which make broader assumptions. These differences need to be taken into account when comparing results.

The distributed borrowing algorithm based on D-LBSB was implemented following the descriptions in Das, Sen and Jayaram (1997) and Das et al. (1997). However, some of the parameters for implementation have not been described fully in the references and a few were changed due to implementation characteristics of the cellular model simulated. Although the scheme implemented generally follows the behaviour of D-LBSB, they are not the same and therefore we refer to our scheme as the *distributed borrowing algorithm* (D-BA). To analyse the performance of D-BA, we chose the system layout with the non-uniform traffic distributions of Zhang and Yum (1989), as shown in Figure 14.1.

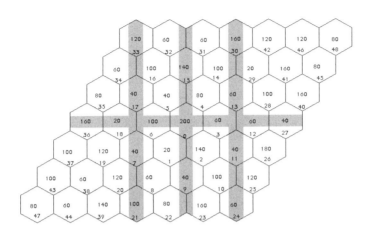

Figure 14.1   Cellular system layout

The number in the bottom of each cell is the cell identification; the number in the middle gives the Poisson arrival rates in calls/hour (ranging from 20 to 200 calls/hour). We also introduced four major trajectories for mobiles (although they can have any kind of trajectory), represented by the shaded areas. Mobile users inside the shaded area can have two speeds: driving at 40 km/h or walking at 2 km/h, and they move in both directions.

The simulation result shows that D-BA outperforms FCA for moderate to heavy load; the maximum improvement achieved in this case is a reduction in call blocking rate by around 20%. However, individual cells with higher traffic rates present a better improvement.

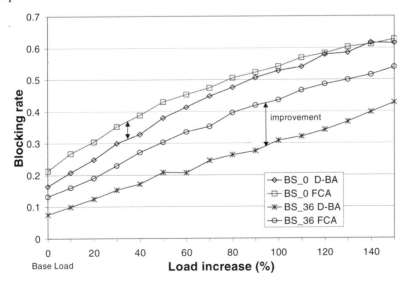

Figure 14.2   Blocking rate of distributed borrowing algorithm versus FCA

Figure 14.2 shows the blocking rate presented by cells 0 and 36 when FCA and D-BA schemes are applied. The abscissa is the percentage of increase over the load shown in Figure 14.1. It is noticeable that as the load of the network increases, the performance of D-BA approaches that of FCAs. Because a heavily loaded network will not have spare channels that can be borrowed, running the algorithm in these traffic conditions represents only overhead for the network. Those fully reactive algorithms proposed in the literature are not able to detect the drop in their efficiency under different traffic conditions. It is clear that in order to avoid wasting signalling resources, base stations must detect these situations and decide on alternative ways of improving the efficiency of the algorithm, or stop it to avoid signalling overhead.

The actual performance of D-BA over different traffic loads is a good parameter for analysis. In D-BA, the borrowing algorithm is not executed every time a call or handoff request is made and there are no more available channels to accommodate the request: it is triggered *before* the nominal channels are all used, once h is reached. Moreover, it does not get only one channel, but a certain number of channels (X) – the actual number depending on the average traffic load of the whole network. In our implementation, each run of the algorithm can be *successful* (all X channels are borrowed), *partially successful* (when some channels are borrowed, but not the number expected), or *unsuccessful*. Failure can be for one of three reasons:

- There were lenders but channel locking was not possible
- All possible lenders are *hot*
- The network is so heavily loaded that X is zero.

If the algorithm has partial success or failure, it will not be allowed to re-execute the algorithm immediately, but it will wait for a certain time depending on the cause of the failure. The introduction of these delays is to avoid continual repetition of the algorithm, which would substantially increase the signalling load to no avail. Table 14.1 shows the results of D-BA attempts during the simulations.

Table 14.1   Results of borrowing attempts

% of load increase	Total no. of D-BA runs	D-BA runs with *success*	D-BA Runs with *partial success*	D-BA *failure*: channel locking not possible	D-BA *failure*: all possible lenders are hot	D-BA *failure*: network is congested (X = 0)
0	6232	2028	1820	1243	1141	0
50	5149	1061	597	1611	1880	0
100	4460	616	45	1344	2406	49
150	3493	206	14	555	1527	1084

The algorithm has good performance at base load, with 38% failure. With an increase in load of 50%, there is a 67% failure rate, rising to 85% at 100% increase. It should also be noted that our policy of introducing a delay reduces the number of attempts to borrow as the load increases, so saving needless signalling messages.

It is clear that a surveillance process in the base station could recognise the conditions when the algorithm is not being efficient and avoid redundant signalling. More detailed analysis of the results shows that up to approximately 100% load increase, the algorithm gets partial success throughout the duration of the simulation run. This means that, overall, resources *are available* to reduce congestion, but the reactive algorithm is not accessing them in an efficient way. To improve the performance and efficiency, we propose a multi-agent system and a particular agent architecture that allows base stations to be more flexible and intelligent, negotiating and cooperating with others to improve the efficiency of the channel assignment scheme. This allows the approach to include *planning* to attempt to balance the load *in advance* of reactive requests.

14.3   The Proposed Scheme Using Intelligent Agents

A multi-agent system (MAS) can be defined as a group of autonomous agents with specific roles in an organisational structure (Müller, 1996). The agents interact with the environment and with each other in a coordinated way, as collaborators or competitors, seeking to fulfil the local or global aims of the organisation. The definition of an agent and the main characteristics that distinguish agents from other software systems can de found in Bodanese and Cuthbert (1999). Agent architectures are classified by the degree of reasoning incorporated by the agent, from a completely logical model (known as Belief, Desire, Intention (BDI) architectures) to a fully reactive model with no symbolic representation. Hybrid architectures combine features of logical and reactive models and are more suitable for real-time applications. INTERRAP (Müller, 1996) is a hybrid agent architecture that also incorporates mechanisms for coordination and cooperation among autonomous agents. It consists of a set of hierarchical layers, a knowledge base that supports the representation of different abstraction levels of knowledge and a well-defined control architecture that ensures coherent interaction among layers. It was designed to react to unexpected events, to long-term actions based on goals and to cope with other agent interactions. INTERRAP was, therefore, chosen as our model for agent implementation. The architecture illustrated in Figure 14.3 was adapted from Müller (1996).

In our scenario, the cellular network has one agent per cell. The *world interface* presented to the agent includes the *sensors section* responsible for the perception of the environment, which would include requests for channel allocation from new calls, handoff requests, borrowing channel requests and orders for locking channels.

The *communication section* handles message exchanges for channel management and in the negotiation process. The *actors section* is responsible for all execution tasks that actually allocate, release, reallocate, lock, lend channels, manage normal and supervised handoffs, and terminate appropriately unsuccessful requests.

In the *knowledge base*, the *world model* contains the environmental information and everything necessary for the operation of the reactive layer. The *mental model* contains complete information about the agent, about the use of frequency channels and possibly a history of traffic load in the cell. Finally the *social model* has relevant information about other agents' data.

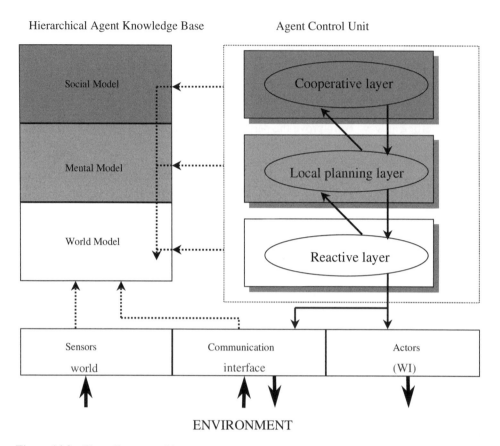

Figure 14.3   The cell agent architecture

The control unit is structured to include a *reactive control layer* that is responsible for fast accommodation of traffic demand, a *local planning control layer* using other strategies to optimise the local load distribution of channels, and the *cooperative control layer*, responsible for load balancing across a larger area.

14.3.1   General description of the agent layers

The function of the control unit of the agent (Figure 14.3) is to recognise a situation from the world interface, activate the goals, plan them, schedule the tasks and execute them. Figure 14.4 shows the flow of the execution cycle of the layer. The situation recognition and goal activation (SG_i) functions are linked in one process, as well as the planning, scheduling and execution (PS_i) functions.

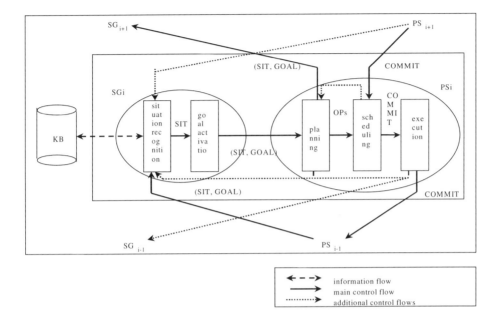

Figure 14.4   Agent control layer

A well-defined control architecture inside a layer, and between layers, is vital for the consistent execution of tasks: in the general cycle description (Figure 14.4), it is shown as interleaved links. There is one link from the *scheduling* function to the *planning* function, when planned operational primitives cannot be scheduled, forcing a re-planning. Another link is from the *execution* function to the *situation recognition* function in the same or previous layer. This is needed because tasks being executed in the environment need to monitor the changes to that environment and the consequences that may arise. Finally, another link exists from the *planning* function to the *situation recognition* function to supply more information, if necessary.

The complete behaviour of the agent is a result of the interplay among control layers. The basic control directions are *upward activation requests* and *downward-acting commitments*. Three control paths can be recognised: reactive path, local planning path and cooperative path.

As can be seen in the representation of the control layer in Figure 14.4 and the control paths in Figure 14.5, the flow of control has two directions. When PS_i decides it is not competent to deal with the SG_i pair, it sends an upward activation request to the next higher layer – the information provided by layer i and the additional information available in layer $i+1$ can produce a suitable goal description for PS_{i+1}. On the other hand, the interaction is organised in a top-down manner. The planning and scheduling processes of neighbouring layers coordinate their activities, communicating commitments in the downward direction, finally causing the execution of actions in the world interface. An acknowledgement protocol is used to monitor success or failure.

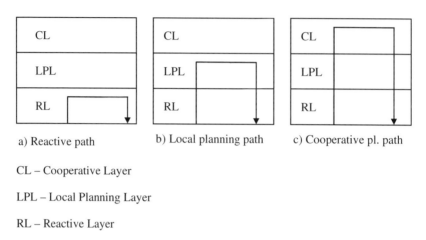

a) Reactive path   b) Local planning path   c) Cooperative pl. path

CL – Cooperative Layer

LPL – Local Planning Layer

RL – Reactive Layer

Figure 14.5   Generic control paths

There is an additional inter-layer coordination: LPL can enable and disable elementary tasks of the reactive layer as a mechanism to control its activity, for example, if the plan has changed. Another possibility is from RL to LPL, to devise a plan for a situation-goal description, to evaluate or interpret a given plan, and to stop activity regarding an earlier request. The cooperative layer can ask LPL to interpret or evaluate a plan, or to stop activity; in the same way, LPL can ask CL to devise a plan given a situation-goal description and to evaluate and interpret a joint plan, perhaps proposed by another agent.

In the implementation of the agent for the cellular scenario, the abstract architecture described so far is embedded inside the control and functional processes that compose the agent. The next subsection emphasises the functionality of each agent layer and how the agents accomplish the complete channel allocation scheme.

14.3.2   Implementation

In the execution of the complete scheme, the agents follow two different strategies: one for channel allocation inside the cell and the other for load balancing of the network. The channel allocation strategy is local and mainly executed by the reactive layer, with optimisation of channel usage being done by the LPL. The second strategy is a joint plan among agents, executed when the local strategy becomes saturated and the performance of the system starts to degrade. The joint plan is triggered by the LPL and coordinated by the CL. The strategies do not conflict, but even so the interaction between layers follows the INTERRAP control architecture, suppressing or changing tasks in the reactive layer when channel allocation conflict arises. The LPL is responsible for keeping consistency between joint plans and reactive tasks.

14.3.2.1 Reactive layer

The reactive layer is basically composed of an FCA algorithm with channel borrowing and channel locking. We decided to use the D-BA implementation in order to measure the benefits of the other two layers of the agent. Below we give a general summary of its implementation.

D-BA deals with three categories of mobile users: *new, departing* and *others. New* users are those that have successfully received a channel in the cell (new calls and handoff requests) between the time of the channel allocation and time *t*. After that time, they will be classified as *others* or *departing* according to their position inside the cell. Mobile users are classified as *departing* if they are inside a determined region (*r*) close to the borders of the cell. In our implementation, the region *r* is fixed, based on the signal strength of the current cell and the neighbouring cells. There is a special routine inside the agent, which checks the position of the mobile user after a certain time to see if the user is really departing or whether it has come back to the main region of the cell. Users are classified as *others* if they are inside the main region of the cell. The routine that classifies mobile users is independent of the borrowing algorithm. The approach taken is to have the user classification continuously updated, allowing the agent to use this information also outside the execution of the algorithm.

When the channel availability in the cell decreases to a certain threshold *h* (for example, 20% as used in the simulation), it becomes 'hot'; cells above the threshold are called 'cold' cells. Hot cells are not allowed to lend channels and cold cells are not allowed to borrow channels. When a cell becomes hot, it triggers the execution of the following algorithm:

- The hot cell broadcasts a message requesting channel availability (called degree of coldness (dc)) from all cells.
- Receiving the responses, it calculates the average channel availability of the entire network (dc_{avg}) and the set of hot cells inside its compact pattern (H_{NCC}). With this data, it calculates how many channels it needs to borrow to get the average channel availability, using the following formula: $X = C \times (dc_{avg} - h)$, where X is the number of channels to be borrowed and C is the number of nominal channels.
- The algorithm starts to execute the cycle below:

1. The hot cell B sends messages to the cold neighbouring cells L. The contents of the message allow the L cells to compute the utility function below. The computations of each L cell are sent back to B.

$$F(B,L) = \frac{dc(L)}{\dfrac{D(B,L) \times (1 + H(B,L))}{Rcp \times 7}}$$

D(B,L) or *nearness* is the measure in terms of cell distance between L and B, for example, for the neighbouring cells of B, D(B,L) is 1.
H(B,L) or *hot-cell channel blockade* (Das et al., 1997) is the number of hot co-channel cells of L that are not co-channel cells of B.

R_{cp} is the radius of the compact pattern, also measured in terms of cell distance. In the formula, R_{cp} and 7 are used for normalisation.

2. B orders the list of L cells by decreasing values of F and informs the cell with the higher value that it is the current lender.
3. B borrows channels from L respecting the safety threshold of L's channel availability. Every channel borrowed is locked in L and in the co-channel cells of L that are not co-channel cells of B inside the reuse distance. If the number of channels needed is not reached, the procedure is repeated for the other L cells in the list. The algorithm terminates when the number of channels X is reached or when the search in the L cells is exhausted.

The reactive layer is responsible for the channel assignment. If the incoming request is a handoff request it will try to assign first to a nominal channel; if that is not possible, it will look for a borrowed channel (if they exist in the cell). If both attempts fail, the request is blocked. If the request is an incoming call, it will look first for an available borrowed channel and then for an available nominal channel. Failure in both attempts leads to the request being blocked.

The mental model keeps the user classification list and the list of departing users updated at all times.

14.3.2.2   Local planning layer

The planning layer determines the departing region based on the signal/noise ratio and traffic history conditions. This layer is responsible for the channel reassignment scheme. Every time a channel is released, the reactive layer requests a reassignment decision:

*   If the channel is a borrowed channel, it will be reallocated to an appropriately departing user (using a nominal channel) or it will be given back to the owner cell, depending on the channel availability of the current cell.
*   If it is a nominal channel, it will be reallocated to a non-departing user that is using a borrowed channel.

The threshold for keeping or releasing a borrowed channel can vary accordingly with traffic prediction or traffic history. Much more intelligence and learning can be added to this layer to improve local performance.

This layer monitors the efficiency of the algorithm in the reactive layer, and it is responsible for the decision of triggering the cooperative layer.

14.3.2.3   Cooperative layer

It is shown in Table 14.1 that in several situations the use of a channel allocation scheme on its own is not sufficient to keep low rates of blocking probability. A well-coordinated joint strategy using supervised handoffs can alleviate the load of a hot spot, moving calls to less loaded regions where possible. This is the responsibility of the cooperative layer of the agents. We are applying a collaborative strategy using the contract-net protocol (CNP) (Smith, 1988). CNP is a classical market-based control coordination technique for task and resource allocation in MAS. The agents can play two different roles: manager or contractor.

When an agent has a task to perform, but is not capable of doing it, the agent becomes a manager and looks for the most suitable contractor by announcing a contract to other agents and selecting the best one according to the bids received. The interesting feature is that a contractor can ask for subcontractors (i.e. become a manager itself) to help it to perform its task. In our case, the agent of the hot cell, using the contract-net protocol, requests its co-channel cells to make offers for moving calls to their regions. To be able to make a bid, the co-channel cells need to compute a utility function F' which requires the load conditions of their compact pattern. Gathering the information from its compact pattern and computing the value of F', it sends this as a bid. The region with the biggest value for F' is chosen by the hot cell. The whole compact pattern of the chosen region starts to ask the mobiles in the more distant departing areas from the originating hot cell to verify the quality of the reception of the control channel of the closest neighbouring cells. If the quality of the reception is good enough the base station will request a supervised handoff for the neighbouring cell with best transmission quality and the mobile will possibly be moved to that cell. The originating hot cell then asks the two neighbouring cells in that direction to start supervised handoffs as well. The execution of the joint plan is carefully coordinated by the market-control technique. The joint plan can be dismissed by the managers when the traffic returns to a value that will allow the borrowing algorithm to be used on its own.

In the CNP we take care to avoid producing a storm of messages: this is the reason that the first possible contractors are the co-channel cells only, which then subcontract their compact pattern.

This layer is currently in the implementation stage the simulation results will allow investigation of the performance and flexibility that can be achieved by the multi-agent system in comparison with a pure reactive approach.

14.4 Conclusions

Resource flexibility is one of the most important requirements in the next generation of mobile communications. Means of increasing the flexibility of the network to deal with new services and traffic characteristics is a requirement and an implementation challenge. This chapter demonstrates the lack of flexibility presented by fully reactive resource allocation schemes by implementing a distributed channel allocation scheme and analysing its simulation results. A more flexible scheme using intelligent agents is proposed. The agent architecture adopted is able to provide greater autonomy to the base stations, which allows cooperation and negotiation among them. A description of the agent implementation is given.

14.5 References

Bodanese, E.L. and Cuthbert, L. (1999) Distributed Channel Allocation Scheme for Cellular Networks using intelligent agents, in *Proceedings of the 7th International Conference in Telecommunication Systems, Nashville TN,* pp. 156–165.

Das, S.K., Sen, S.K. and Jayaram, R. (1997) A Dynamic Load Balancing Strategy for Channel Assignment Using Selective Borrowing in Cellular Mobile Environment. *Wireless Networks*, **3**, 333–347.

Das, S.K., Sen, S.K., Jayaram, R. and Agrawal, P. (1997) A Distributed Load Balancing Algorithm for the Hot Cell Problem in Cellular Mobile Networks, in *Proceedings of the 6th IEEE International Symposium on High Performance Distributed Computing, Portland, USA*.

Elnoubi , S.M., Singh, R. and Gupta, S.C. (1982) A New Frequency Channel Assignment Algorithm in High Capacity Mobile Communication Systems. *IEEE Transactions on Vehicular Technology*, **21(3)**, pp. 125–131.

Jiang, H. and Rappaport, S.S. (1994) CBWL: a New Channel Assignment and Sharing Method for Cellular Communication Systems. *IEEE Transactions on Vehicular Technology*, **43(2)**.

Karlsson, J. and Eklundh, B. (1999) A Cellular Mobile Telephone System with Load Sharing – an Enhancement of Directed Retry. *IEEE Transactions on Communications*, **37(5)**.

Katzela, I. and Naghshineh, M. (1996) Channel Assignment Schemes for Cellular Mobile Telecommunication Systems: a Comprehensive Survey. *IEEE Personal Communications Magazine*, **3(3)**, 10–31.

MacDonald, V.H. (1979) Advanced Mobile Phone Services: the Cellular Concept. *Bell Systems Technology Journal (Special Issue)*, **58(1)**.

Müller, J.P. (1996) The Design of Intelligent Agents: A layered Approach, in J. Carbonnel and J. Siekmann (Eds.) *Lecture Notes in Artificial Intelligence 1177*. Springer.

Smith, R.G. (1988) The Contract Net Protocol: High-Level Communication and Control in a Distributed Problem Solver, in A. Bond and L. Gasser (Eds.) *Readings in Distributed Artificial Intelligence*. Morgan Kaufman.

Zhang, M. and Yum, T. S. P. (1989) Comparisons of Channel Assignment Strategies in Cellular Mobile Telephone Systems. *IEEE Transactions on Vehicular Technology*, **38(4)**.

15

Predicting Quality-of-Service for Nomadic Applications Using Intelligent Agents

P. Misikangas, M. Mäkelä, K. Raatikainen

15.1 Introduction

Software systems that are to be used in wireless environments should be able to adapt to sudden changes in the quality of data transmission over wireless connections. As a minimum, a system should detect when current data transmission tasks may not be completed any longer in a reasonable amount of time due to temporary changes in the Quality-of-Service (QoS). More sophisticated systems could try to adapt to the current QoS by using special data filtering and compression methods, and to refuse to accept requests that cannot be fulfilled within a certain time limit. A good example is a Web browser that automatically shrinks or ignores large images on the requested Web pages when the QoS is not good enough. However, quite often an adaptation which is started right after a change in QoS is detected comes too late. This is especially true when the connectivity was just lost – nothing can be done after detection of a dropout, but something could have been done beforehand, if the system had been able to predict the dropout.

Predicting changes in the QoS of wireless links will be a fundamental requirement of future systems that are supposed to do intelligent adaptation in wireless environments. Estimates of future QoS can be used, for example, in *scheduling decisions* (e.g. which tasks are allowed to use bandwidth when the connection is about to be lost), *data prefetching* (e.g. downloading something beforehand while the connection is still good), and *connection management* (e.g. closing the connection now to save expenses because the QoS will be inferior during the next five minutes). We believe that predictions of useful accuracy can be made by learning how quantities like *time of day*, *day of week*, *past QoS values*, and *location of the terminal* affect the QoS.

This chapter is structured as follows. First, in section 15.2, we will give a brief overview of the Monads project (Raatikainen, Hippeläinen, Laamanen and Turunen, 1999) in which this research is being conducted. In section 15.3, we will describe the methods we have examined for predicting QoS in our prototype system; since the project is targeted at nomadic users, our main focus will be on how the location of the terminal affects the QoS. The methods have been implemented into intelligent and learning agents that are described in section 15.4. In section 15.5, we will present some preliminary test results obtained from a simulation. In section 15.6, we will discuss issues related to the QoS prediction problem. Finally, we will draw conclusions about the usefulness of our approach and outline future plans for our research.

15.2   Monads Overview

The research project Monads (Raatikainen et al., 1999) examines adaptation agents for nomadic users. Figure 15.1 outlines the native computing and communication environments in Monads. The key elements are terminals, access nodes and service nodes. In the Monads architecture, a mobile terminal device is connected to the fixed network through a weak connection, such as GSM Data (Mouly and Pautet, 1992), GPRS (Brasche and Walke, 1997), or Wireless LAN (Pahlavan, Zahedi and Krisnamurthy, 1997). The access node is a fixed host in the fixed network, which provides connectivity for mobile terminals to the fixed network. An access node can be hosted by a public service provider, or it can be located in the private network of an organisation. The service nodes are hosts in fixed networks providing different kinds of services to both nomadic users and users using wired connections.

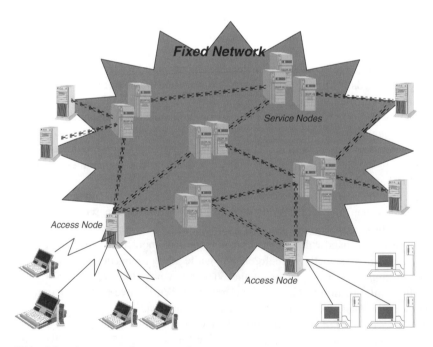

Figure 15.1   Monads system reference configuration

The goal of Monads is to help nomadic users and nomadic application developers in the following ways:

- Improve the efficiency of existing network applications such as Web browsers; this should be possible with zero or minor modifications to these applications.
- Provide a basis for building Monads-specific applications that take full advantage of the Monads system, and therefore are able to use wireless links more sensibly than regular applications.
- Optimise the establishment, configuration and maintenance of wireless connections with regard to cost and response time. Often connections are quite expensive, and in those cases the system should carefully consider whether the connection is really needed, and work off-line whenever possible.

The Monads system is based on the idea of adaptive, collaborative agents, as depicted in Figure 15.2. Each type of agent encapsulates knowledge about its particular domain: data communication agents understand the properties of different communication infrastructures, such as GSM Data, WLAN or GPRS; user interface agents know the capabilities of the terminal, such as its display type; and finally, service agents are aware of the constraints that apply to their service, such as the minimum bandwidth it needs, and know if and how it can be scaled down for low-bandwidth connections.

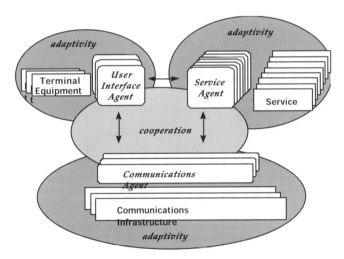

Figure 15.2 Monads adaptation model

Each of these agent types uses its knowledge for adaptation, both internally and collaboratively. Data communication agents adapt to the communication infrastructure, so that service agents do not have to concern themselves with how QoS constraints are mapped to the parameters understood by the communication infrastructure. User interface agents adapt to the capabilities of the terminal device, so that services do not have to concern themselves with the capabilities of the terminal. However, if the service is to operate efficiently, service agents themselves must also adapt. There is no sense in trying to send a operate efficiently, service agents themselves must also adapt. There is no sense in trying to send a high-resolution real-time video over a link that does not have enough bandwidth, for example. Instead, the video should be scaled down, and the frame rate lowered, so that transfer is possible. A service agent can also improve performance if it understands terminal capabilities. For instance, if a terminal cannot show colour images, transferring them in colour to the terminal is inefficient, even if the user interface agent is able to adapt by converting the images to monochrome. Instead, the service agent should convert the images prior to transfer, so that bandwidth is not wasted.

Predicting future QoS is part of the adaptation process. QoS predictions are provided by communication agents, possibly with the help of some service agents such as the location service (see section 15.4). Service agents can then use the predictions to adapt their behaviour to the forthcoming QoS.

15.3   Methods for QoS Prediction

We have divided the original problem of QoS prediction into two subproblems: *predicting terminal movement* (section 15.3.1) and *predicting QoS at a given location* (section 15.3.2). In section 15.3.3, we will describe how these predictions can be combined to answer questions such as 'What is the expected amount of data we can transfer within the next t seconds?', 'What is the expected time to transfer x kilobytes of data?', and 'What is the probability that we can successfully transfer x kilobytes of data within the next t seconds?'. The methods described in the following subsections assume that the underlying probability distribution remains static. In practice, distributions may vary according to time. We handle this by using separate models for different times.

15.3.1 Predicting terminal movement

In order to predict terminal movement, we must first try to learn the movement patterns of the user who carries that terminal. Luckily, most users do not move randomly with their terminals. Instead, they probably use only a few routes in their everyday life – from home to work, from work to home, and so on. Thus, we believe that the movement patterns of a user are relatively easy to learn. Of course, in order to learn anything about terminal movement, we must somehow obtain the current location of the terminal. Hence, from now on we assume that the terminal has some kind of positioning device (e.g. GPS) attached to it.

The concept we would like to learn – regular routes of the user – is hidden in a continuous stream of information about the location and speed of the terminal, received from the positioning device. Therefore, the sequence of coordinates must first be transformed into a much shorter sequence of important *waypoints*. Waypoints are added to locations where the direction of movement or speed changes significantly or routes cross. In the example of Figure 15.3, the original route drawn with a solid line is transformed into a waypoint sequence $\{A; B; C; D; E; F; C; G\}$. We say that a waypoint w_i has a *connection* to w_j if there has been a sequence of waypoints in which w_j came right after w_i. Thus, waypoints and connections between them form a directed graph which we call a *waypoint map*.

Now we can define a *movement pattern* which is a sequence of waypoints that appear frequently. The original problem of location prediction can also be mapped to a problem: given a list of previous waypoints, what will be the following waypoints? At first sight, this problem looks similar to discovering frequent episodes from a sequence of events, which is currently a hot topic in the knowledge discovery field (see e.g. Mannila and Toivonen, 1996). Fortunately, the fact that each waypoint has only a few connections makes this problem a much easier one to solve. In addition, we do not need to use all existing waypoints in the movement prediction task. It is enough to take into account only waypoints from which there are at least two different ways to continue, and waypoints where the user has remained for a long time. Thus, the number of necessary waypoints is so low that we can use a simple and efficient data structure called a *backward tree* (Bell, Cleary and Witten, 1990) for learning and predicting the movement of a user. With a backward tree, we can easily calculate probabilities for every possible path following the given sequence of waypoints.

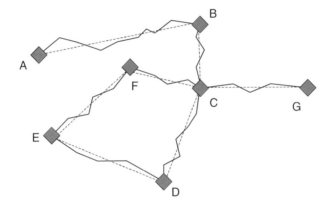

Figure 15.3   Approximating a route with few waypoints

15.3.2   Predicting QoS at a given location

In this section, we concentrate on how to predict QoS in cases where the location is known. We will focus on predicting throughput, but the same methods can also be used to predict other characteristics of QoS like latency and dropout probability. Our approach is based on routes, which gives us a reasonable balance between efficiency and robust predictions.

If we are travelling along a route from which we have collected information about QoS sometime before, a large part of the prediction problem can be described as estimating how confident we are that the values we have observed are a representative sample of the underlying probability distribution. When there are many attributes (e.g. time and location), we must also estimate how much each attribute contributes to the observed values.

We have several ways of saving information about our observations on the route and how this information is used for predictions. For practical reasons, we divide the route between two waypoints into parts where, in each part, the QoS is imagined to be the same from start to finish. This is not a very restrictive assumption so long as that the parts are short enough. Perhaps the simplest way of information saving would be to calculate the mean of all QoS values over the time for each part of the route; however, there are weaknesses in this approach. First, it is static in nature, meaning that even very old observations are as important as new ones; secondly, we somehow need to smoothen the predictions before our sample size is so large that the calculated mean approaches 'the true mean' (i.e. the value given by the underlying probability distribution). The first problem can be solved by, for example, using a cumulative weighted average instead of an average; the second by initialising route-parts with prior values, such as the recent global QoS values. A step forward from using averages, involving a little more calculation, is to use simple probability distributions. Utilising Bayesian methods (see e.g., Gelman, Carlin, Stern and Rubin, 1995), we can then use prior distributions for smoothing and make direct probability statements of QoS within the route-part in question.

So far we have concentrated entirely on predicting QoS on routes the user has travelled previously. To improve initial QoS prediction accuracy and speed up the learning, we can use information gathered from field-tests or from other users. This information can be used to form some kind of *QoS-map* as a prior knowledge of the area's QoS properties, for example. Even when we have some kind of QoS-map of the area, prediction accuracy on the previously unseen route will unavoidably be lower than on a known route. This is due to the fact that even if the route-points could somehow be predicted well (for example, based on the current speed and direction), a two-dimensional QoS-map cannot in general have as much accuracy as a one-dimensional 'route-map'.

There are many reasonable options to implement a QoS-map. In the case of areas with relatively static QoS, a point-based map formed with *vector quantisation* (see e.g. Nasrabadi and King, 1998) is a good candidate. Such a map is efficient to transfer, calculations based on it do not require much processor time, and, further, it is possible to obtain an estimate of the prediction accuracy. However, for efficiency reasons, it is wise to do the calculations for each route-part from the QoS-map only once, so that the map affects the prior distribution of the part.

15.3.3 Combining the predictions

In previous subsections we have looked at ways to calculate probabilities for the user's future routes and QoS on them. Now we must combine the two prediction-sets in order to answer their questions posed earlier. Figure 15.4 clarifies the principles of prediction combining. In this example, the terminal is currently on waypoint A and is predicted to move to waypoint B. From B, we have two alternatives: with 70% certainty, the terminal will continue to waypoint C, but there is also a 30% chance that it will move to waypoint D. From previous experience, the system has learned how some QoS parameter, say throughput, changes while moving from one waypoint to another. To keep this example simple, we use only the average of the QoS parameter. By combining the route prediction and knowledge about the QoS on those routes, we can form a prediction of the QoS in the near future. We know what the QoS will be in the worst and the best case at a given time, and can also calculate a weighted average using the probabilities of the routes.

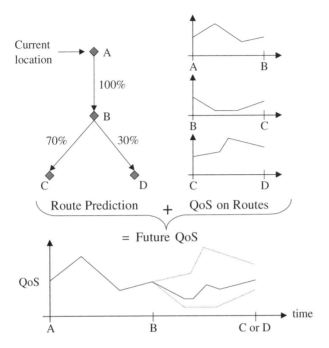

Figure 15.4   Combining predictions

The actual calculations are question-dependent. However, for most questions concerning probability, we must calculate the sum distribution of the QoS aspect in question. For each possible route, we multiply the sum distribution of the involved route-parts with the probability of the route and add to the sum over all possible routes. Now, to answer, for example, the question 'What is the probability that we can successfully transfer X kilobytes of data in Y minutes?', we calculate the portion of the (throughput) sum distribution greater than X. If we used averages for route-parts, we would only get the mean of the expected QoS and must use a separate model to estimate the reliability of the prediction.

15.4   Monads System Components

We have implemented the methods described in section 15.3 into a prototype of the Monads project. The components of the Monads system architecture that involve QoS prediction are shown in Figure 15.5. These components are:

- *Perception Service* is a centralised collection of perceptions. By perception we mean anything that can be observed, for example, location and throughput. The Perception Service takes care of collecting values of perceptions within certain time intervals and storing the values. This centralised handling of perceptions offers many advantages to cope with the problems mentioned in the previous section. Deciding which part of the collected data should be flushed, and when, is much easier to do reasonably for one entity than if all agents invent themselves when it is time to get rid of some data (before storage runs out). Further, finding initially unknown correlations is more likely with this approach, refining data (e.g. clustering) or avoiding overlapping data within several agents is easier, and so forth.

- *Location Service* reads location information from a positioning device (e.g. GPS) and builds/updates a waypoint map according to the terminal movement. Information about the terminal location and waypoints reached are stored in the Perception Service.

- *QoS Management* provides information about the current QoS and stores it in the Perception Service.

- *Route Modeller Agent* is an autonomous agent that tries to learn the regular routes of the user, and provides predictions about future locations of the terminal. It collects data (sequences of reached waypoints) with the help of the Perception Service, and builds/updates the movement model as described in section 15.3.1.

- *QoS Modeller Agent* is an autonomous agent that tries to learn the QoS as a function of location and time, and provides estimates about future QoS at a given location and time. It collects data (time, location, QoS) with the help of the Perception Service, and builds/updates the QoS model as described in section 15.3.2.

- *QoS Prediction Agent* is an intelligent agent that provides predictions about the future QoS by combining predictions made by the modeller agents described above (see section 15.3.3). If there are alternative ways to predict movement and QoS, this agent is responsible for choosing between them. It may also warn other agents when the QoS is about to change significantly.

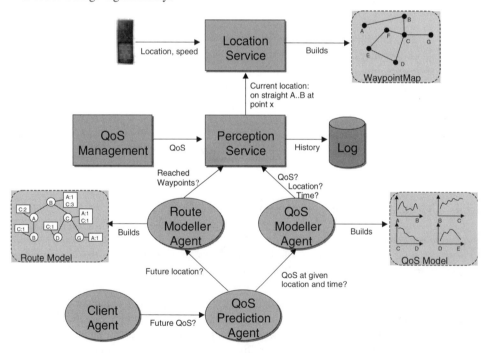

Figure 15.5   Monads system architecture

15.5   Test Results

Our QoS prediction approach is based on an assumption that the expected throughput varies according to location and time. This is not quite true for a GSM data connection[1], but the forthcoming GPRS technology should have this property. In order to test the usefulness of our methods before GPRS is available, we have built some simulation capabilities into the Monads system. Because of the modular design of our system, it was enough to replace the real *Location Service* and *QoS Management Service* with simulated ones called *Route Simulator* and *Throughput Simulator*. The Route Simulator simulates the movement of a user by using a set of predefined waypoints and routes read from a file. The Throughput Simulator produces simulated throughput values for each location visited by using a 'QoS-map' read from a file. Both simulators include randomness, so that each simulation run is different.

One of the questions we asked our QoS Prediction Service was 'What is the expected amount of data that we can transfer within the next *t* units of simulation time?'. For each prediction, we also checked what should have been the correct answer by collecting throughput values for the period in question. We compared two implementations of this service. The first version, called *Nomadic QoS Prediction*, uses the methods we have described in this chapter, i.e. its output is based on predictions of the user movement. The second version bases its predictions only on the average of recent throughput values. We call this latter version *Static QoS Prediction* because it assumes that the throughput remains at the same level as before.

Figure 15.6 shows a typical example of results from a simulation run. The thickest line labelled 'Real' is the correct answer, i.e. the amount of data that actually could have been transferred within the given time limit, and the two other lines are the predictions made by Nomadic/Static QoS Prediction. As expected, the static version performs quite poorly, because it cannot foresee changes in QoS. Thus, if the throughput level changes significantly immediately after a prediction, the prediction may be totally wrong. In the beginning, the Nomadic QoS Prediction behaves just as badly as the static one because the 'user' is moving via unseen routes[2]. However, when on a route travelled previously, the predictions are very close to the correct answer. Figure 15.7 shows the average error made by the methods over several simulation runs with different configurations.

---

[1] In our field tests, most of the time throughput stayed at a very constant level and dropped only for short moments while changing the cell. The only useful thing that we could learn about GSM data connections is the probability of dropouts in different areas, but we have not yet paid much attention to this issue.

[2] No prior knowledge about the QoS-map or user routes was used in the simulation.

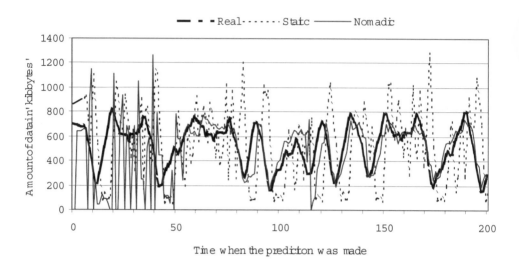

Figure 15.6   Nomadic versus static QoS prediction

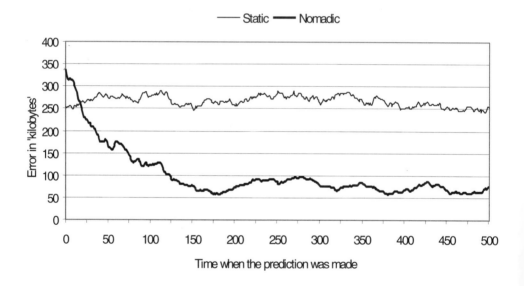

Figure 15.7   Average error

15.6   Discussion

QoS prediction forms a very good benchmark problem for our goal of producing an agent platform supporting intelligent and learning agents and adaptation in wireless (low-bandwidth) systems. It includes many common learning problems, which must be taken into account in order to achieve the best results:

- *Too little information*: it is clear that observations from one terminal are not initially enough for very accurate predictions. Thus, knowledge sharing between different terminals and users becomes an issue. In order to minimise the cost of information sharing over a wireless link, the system should be able to decide which information is worth sending and when.
- *Too much information*: a GPS device connected to the system is a clear example of a case in which some of the information must be flushed or refined in time to prevent storage problems.
- *Missing information*: the models we use can depend on information which may not always be obtainable. Therefore, in order to obtain a robust system, we prefer models that can work with partial information and/or use a set of (hierarchically organised) models to cope satisfactorily with different kinds of situations.
- *Unknown correlations*: sometimes we could, in principle, have the information to predict a change in QoS, but we do not know beforehand that the information in question correlates with future QoS values. This kind of information could be produced, for example, by an agent which buys flight tickets for the user; it is very likely that the connection will go down sometime before the beginning of the flight and is not going to resume before the arrival.

In addition, the problem can be disassembled into subproblems as explained in section 15.3. However, there is no obvious best solution, covering all situations, for dividing the components across the network. In particular, we must decide which components should reside in the terminal and which should be placed into the access node. In general, components should be near to the source of the information they need, so that the information does not need to be sent over the wireless link. Unfortunately, it is usually impossible to have all components on the same side, and the optimal configuration may vary according to the situation. Therefore the system should be able to adapt to the changing network conditions, and reorganise the placement of components if necessary.

It is clear that to cope with all of the above problems requires a very modular approach, especially because we are talking about nomadic users and thus the scalability of the system is of major concern. The system should also be dynamic in the sense that parts of it are exchangeable to adapt to, for example, changing hardware in terminals. Considering these problems and requirements has led us to believe that agents can indeed be of valuable help in the prediction of QoS.

15.7 Conclusions

We have introduced a way of predicting available Quality-of-Service in wireless networks using intelligent agents that learn movement patterns of a user and the characteristics of available throughput in space. In the near future, as well as throughput we will also be considering other aspects of QoS. The reported results clearly indicate that agents can quite quickly learn characteristics that enable reasonable short-term predictions.

The flexibility of the Monads agent system allows the introduction of various kinds of modeller agents, each of which concentrates on learning a specific aspect of user or network behaviour. These agents utilise the Monads system services and provide predictions to agents which can then combine those predictions in order to give predictions at higher levels of abstraction. In the future we will develop additional modeller agents and concentrate on the adaptation problems introduced by the need of information sharing between modeller agents. We will also examine alternative learning algorithms and pay special attention to combining predictions.

Acknowledgements

This work was carried out as a part of the Monads research project funded by Sonera Ltd, Nokia, and the National Technology Agency of Finland. The authors express their thanks to the rest of the Monads team.

15.8   References

Bell, T., Cleary, J. and Witten, I. (1990) *Text Compression*. Prentice Hall, Englewood Cliffs, N.J.

Brasche, G. and Walke, B. (1997) Concepts, Services, and Protocols of the New GSM Phase 2+ General Packet Radio Service. *IEEE Communications Magazine,* **35(8),** 94–104.

Gelman, A., Carlin, J., Stern, H. and Rubin, D. (1995) *Bayesian Data Analysis*. Chapman & Hall.

Mannila, H. and Toivonen, H. (1996) Discovering Generalised Episodes Using Minimal Occurrences, in *Proceedings of the 2nd International Conference on Knowledge Discovery and Data Mining*, pp. 146–151, AAAI Press.

Mouly, M. and Pautet, M.-B. (1992) *The GSM System for Mobile Communications*.

Nasrabadi, N. and King, R. (1998) Image Coding Using Vector Quantisation: A Review. *IEEE Transactions on Communications*, **36(8)**, 957–971.

Pahlavan, K., Zahedi, A. and Krisnamurthy, P. (1997) Wideband Local Access: Wireless LAN and Wireless ATM. *IEEE Communications Magazine,* **35(11)**, 34–40.

Raatikainen, K., Hippeläinen, L., Laamanen, H. and Turunen, M. (1999) Monads – Adaptation Agents for Nomadic Users, in *Proceedings of the Infrastructure Forum in Telecom 99*, Geneva, Switzerland, ITU.

16

Implementation of Mobile Agents for WDM Network Management: The OPTIMA Perspective

D. Rossier-Ramuz, R. Scheurer

16.1   Introduction

In the near future, the emergence of new photonic technologies in the field of transport networks, such as WDM (Wavelength Division Multiplexing) optical networks, will support a variety of services such as voice, audio, video and Internet traffic. Today, one single optical fibre can transport up to 320 Gbit/s and the capacity of more than 1 Tbit/s is expected within about five years. The incredible growth of the bandwidth through one fibre, as well as future IP/WDM mesh networks, requires the implementation of some efficient mechanisms for the management of such networks, especially concerning reaction speed in case of fibre break or component damage.

During the last few years, we have observed a significant convergence between Information Technology and Telecommunications. New open distributed systems have offered network management a large set of usable solutions. For example, the world-wide TMN manager/agent environment (ITU, 1996) provides a generalised framework for the management of telecommunication networks. It defines a kind of agent – hereafter called *stationary* agent – that can interact with a manager. This platform-centred approach, however, entails drawbacks in scalability, reliability, efficiency and flexibility, and is consequently unsuitable for large and heterogeneous networks (Bieszczad, Pagurek and White, 1998). Future optical networks, for example, which will carry any other network layer or service (e.g. IP/WDM, SDH/WDM), thus making progress in the direction of a transparent WDM network, will require a fully managed WDM network layer (Lagasse et al., 1998).

Recent work in the field of distributed artificial intelligence, in particular towards distributed *stigmergetic*[1] control issued by societies of biologically inspired agents, provides a wide range of mobile-agent-based solutions in order to solve problems related to network management (Bieszczad et al., 1998; Di Caro and Dorigo, 1998). However, a lack of suitable infrastructure for the support of autonomous agent-based control, security issues for mobile agents, as well as interconnection with legacy systems, do not favour agent technology in the telecommunication world (Hayzelden and Bigham, 1999). For this reason, a need exists for establishing proven solutions to show that it is possible to implement agents efficiently into network infrastructure. From the same point of view, it seems important to observe and understand the current standardisation work carried out by organisations such as ITU[2], OMG[3], or TMF[4], and to convince the telecommunications world that an agents-based approach can match their expectations.

In short, this chapter outlines ongoing research activities in the field of multi-layer optical network management based on a new concept of so-called 'optical' agents. In this context, we will be involved in the OPTIMA[5] project. The objective of this project is the definition and elaboration of a new intelligent optical platform whose suggested approach is to implement optical agents within the network elements themselves, or within close proximity to them.

16.2   Future IP/WDM Networks

The current trend in the networking world is the development of a paradigm that enables the transport of layer 3 traffic (IP) over optical networks. Still, cost-effective transport solutions based on IP/WDM networks are only attractive if they can deal with legacy systems like SDH/WDM or ATM/SDH/WDM. We must therefore consider that IP/WDM and SDH/WDM are to co-exist on the same optical infrastructure in order to ensure efficient migration. Major issues concerning future service-independent optical networks are related to their management and, especially, to fault management. In the context of a TMN environment, the information flow between managers and agents increases, and time for making decisions depending on human intervention is becoming a critical issue. The introduction of intelligent and mobile agents within optical networks should lead to the reduction of the information flow through the management information channels and, moreover, should spare the usage of numerous protocol stacks (the solutions proposed by different vendors are often incompatible and make the integration of heterogeneous networks very difficult). Owing to this approach, optical networks will become more

---

[1] Stigmergy is defined as 'the indirect communication taking place among individuals through modifications induced in their environments'.

[2] International Telecommunication Union. Web site: http://www.itu.int [06.06.2000].

[3] Object Management Group. Web site: http://www.omg.org [06.06.2000].

[4] TeleManagement Forum. Web site: http://www.tmforum.com [06.06.200].

[5] OPTIMA - OPTical network with Intelligent and Mobile Agents - A Swiss National Science Foundation project under grants 21-61922.00

generic: they will be able to support the future services requirements and to guarantee a fast reaction to the dynamic changes in traffic conditions; automatic reconfiguration should reduce human intervention and with it the risk of errors as well as the reaction time. The development of this approach constitutes a major objective in the OPTIMA project.

16.3   Optical Transport Network

16.3.1   Introduction

Optical Transport Network (OTN) is another name for an optical WDM network, which is currently being standardised by the G.872 recommendation of ITU-T (ITU, 1999). It is based on WDM technology, which allows multiplexing of several wavelengths and leads to a tremendous increase in available bandwidth through a single optical fibre. These future networks open a new field of interesting issues related to wavelength handling (provisioning, (re-)routing, etc.), which can be influenced by a variety of logical and physical parameters. OTN is defined by three sublayers, which are the Optical Channel layer (OCh layer), the Optical Multiplexing Section layer (OMS layer) and the Optical Transport Section layer (OTS layer).

We can briefly describe the OCh layer as a server layer on which any kind of rate-independent clients (e.g. IP, SDH, ATM) can initiate connections between two nodes. A client is associated to one wavelength (or channel). Additional specific parameters related to the client demand, such as Quality of Service (IPv6 type of service (e.g. Huitema, 1998)) or protection paths for an SDH client (ITU, 1998), have to be taken into consideration so that interoperability issues can be solved and the client signal can be mapped on adequate wavelength frequencies according to local constraints. Actually, if the same wavelength is assigned to the connections on all the links, the blocking probability increases considerably. In particular, the OCh layer contains a wavelength routing function, while the OMS layer is responsible for ensuring integrity of a group of wavelengths (wavelength multiplex) and the OTS layer is the interface between OMS and physical layers.

16.3.2   Transport of agents

It is most probable that a special frame structure, such as a digital wrapper, as recently proposed within ITU by Lucent Technology (1999), will be implemented within the OCh layer of OTN in order to support, for each channel, an overhead, a payload at different bit rates and an additional FEC (Forward Error Correction). The overhead, which is processed by a special atomic function in the OCh layer, can support bytes reserved for management purposes, such as the Embedded Communication Channel (ECC). At the OMS layer, another communication channel is defined for the support of management information related to this layer; it is called the Optical Supervisor Channel (OSC). OSC requires a specific wavelength. Figure 16.1 shows the different layers, the OSC and the frame structure of a digital wrapper (functional model based on ITU (2000)).

Digital wrapper and OSC both offer an interesting way to allow an ECC to be transported within the optical network. The ECC could transport agents' code with restrictions in order to keep the code size small and to reduce the amount of communication with other agents. We intend to use these two communication channels in order to transport a new family of agents called *optical agents* (or λ-*agents*). Basically, these small agents will evolve within the network using little or no communication with other agents. The separation of payload from overhead provides the agent transportation with a security mechanism, since the overhead processing is completely carried out within the network element and has no interaction with the payload. As we will see further on, we have defined two kinds of agent: λc-agents and λm-agents (Figure 16.1).

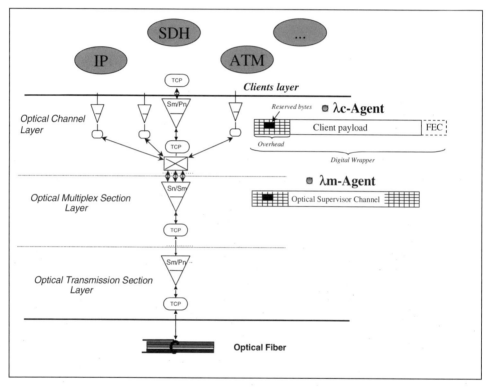

Figure 16.1   Overview of OTN layers and the communication means available for management
purposes

Details concerning the structure of the λ-agents, which are currently being studied, lie beyond the scope of this paper. λ-agents are characterised by an interesting property: they are able to move rapidly from one node to another with the possibility of being routed very fast through the network nodes (multi-hop capability). Overhead processing, which is performed by the atomic function of the OCh layer, could actually be extended, so that the dedicated bytes are passed to an appropriate agent transport function. This function will retrieve the destination address of the λ-agent, thus allowing the agent to be forwarded to the next node without using the resident mobile agent framework. This property facilitates the assignment of a priority in the transport of our agents.

It is also fundamental that the agents are transported in the best conditions, without any transfer errors. From this point of view, the application of FEC to the overhead should guarantee better reliability.

16.3.3 Fault management in OTN

As several Tbit/s are expected to be transported through only one optical fibre in the near future, fault protection and service restoration are highly critical; the former consists of finding disjoint protection paths in order to re-route the traffic in case of failure; the latter consists of switching the broken paths to the precomputed protection, as rapidly as possible, in order to restore the traffic conditions prevailing before the failure. Such issues lead to various analytical formulations of the all-optical network design problems, for which various heuristic algorithms have been developed (Yoo, 1997). While the service restoration process is running, a fault location process is started to locate the failure as precisely as possible. In the scope of our research, we have decided to focus on two major aspects of fault management: fault protection and fault location.

Fault protection looks for optimal protection paths based on client-layer requirements and on physical parameters, laser temperature, for example. Physical parameters give information concerning potential failures, also called progressive failures, and allow avoidance of certain paths in favour of others. When a failure occurs, the traffic is re-routed to the protection paths, using very fast switching mechanisms implemented within the optical components. As the evolution of traffic demands is getting more and more dynamic – consider Internet connections as well as future STM-n on-demand in the case of SONET/SDH – it is absolutely essential to implement dynamic mechanisms in order to recompute new optimal protection paths according to traffic conditions at a given time. Fault protection starts as soon as a client is asking for an end-to-end connection (provisioning process).

Fault location consists of locating the network element (NE) that is responsible for malfunction. In case of fibre cut, we will try to find the NEs close to the damaged fibre. The fault location process therefore starts just after the failure. An algorithm based on a centralised architecture and filtering of messages received by the manager has been proposed in Mas and Le Boudec (1997).

16.3.3.1 Reaction time issues

As with typical problems concerning transport-layer integration, we can mention the integration of SDH rings network on mesh WDM network, which leads to serious problems of interoperability regarding the protection strategy; three distinct problems have been identified by Crochat (1998): bottleneck, connectivity and multi-group problems. These typically NP-complete problems can be solved by resorting to specific heuristics but, as this requires considerable processing time, the algorithms are executed during the provisioning process in a centralised manager. The network elements are eventually set up with the precomputed paths. Another solution consists of using dynamic re-routing after a failure but, once again, this process may require considerable time and may not ensure maximum failure survivability.

Instead of using algorithms which do not take traffic evolution into account, we propose to let the λ-agents find protection paths and configure the network elements according to their travel results.

16.3.3.2 Dynamic re-routing mechanism

The optical components also contain special stationary agents called switching-protection agents (Sexton and Reid, 1997), which are implemented inside the elements and enable the equipment to initiate switching to protection paths in case of failure conditions. These agents can be considered as TMN-like agents and are able to access managed objects defined in the MIB component which provide information concerning the configuration of protection paths.

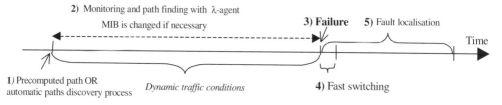

Figure 16.2   Example of dynamics with λ-agents keeping MIB up-to-date

As already mentioned, the dynamic characteristics of demands addressed to the OTN layer require a continuous validation of protection paths; it may also be necessary to update the protection paths under normal traffic conditions. We propose to introduce a dynamic re-routing precomputation mechanism (Figure 16.2), which consists of launching optical agents into the network and letting them try to find new paths according to current service requirements. Optical agents exchange information with stationary agents, which keep the local MIB of the elements concerned up-to-date. This approach is similar to a dynamic re-routing but, at failure time, switching-protection agents have already identified protection paths and can immediately initiate fast hardware switching.

Quick convergence towards a solution, which remains to be studied, will ensure that protection paths are based on constraints corresponding to the traffic conditions established before the failure.

16.4   Optical Agents

16.4.1   Introduction

The paradigms of intelligent and mobile agents are already used world-wide with various meanings. We therefore propose to clarify the terms: agent, intelligent, mobile and then optical in our context.

The following four common properties appear from the various definitions of an agent (Wooldridge and Jennings, 1995): *autonomy, social ability, reactivity* and *pro-activeness*. In our context, the agent in general is small and should therefore perform only simple tasks (individual ants are unsophisticated insects). Let us now examine the four above-mentioned properties in detail, in the context of our approach.

Autonomy contributes to reducing the information flow between normal managers and agents, as implemented within a TMN-like environment; it allows enhancement of the network element by providing it with more intelligence. The agents have their own execution thread and could communicate with other agents or with an agency in order to exchange information. The agency is responsible for managing a subset of agents and surveying their activities. Optical path discovery, path provisioning and fault location can be rapidly performed within the transport network, without requiring the intervention of other entities – or human intervention – at upper layers. Performing such tasks requires the usage of several agents spread over the nodes of the network.

Studies of the interactions between the agents themselves, or between the agents and their environment, refer to the second characteristic: *social ability*. While we promote agent societies with small code size and little communication between agents, social ability[6] refers to the ability of an agent society to seek a common goal, although the agent itself has no knowledge of the ultimate goal. The simulation of several kinds of agent societies will allow us to determine the degree of interaction between societies.

Reactivity implies that agents are aware of their environment and can react to changes occurring in it. In our context, the environment is characterised by two groups of factors: possibly complex client requirements, such as protection strategy at the client layer, quality of services or other provisioning constraints, or events related to a failure (e.g. fibre cut, equipment defect) and parameters related to progressive failure. Both these factors are taken into account by different specific λ-agents: λ-agents for fault protection or λ-agents for fault location.

Pro-activeness refers to the faculty of making decisions (such as leaving one node for another one, or suspending activities). This property is actually implied in reactivity: in the case of fibre cut, an agent detects that no traffic is available on a specific wavelength and assigns a λ-agent to fault location, while another λ-agent starts looking for new protection paths, if any protection path was attached to the defect channel.

From our point of view, the intelligence paradigm is related to the expression of intelligent behaviour in agent societies. This part of artificial intelligence is concerned with multi-agent systems and is based on principles inspired by biological behaviour, such as the behaviour of ant societies. Di Caro and Dorigo (1998) have studied the use of an algorithm based on *ant colony optimisation* and applied it to routing in packet-switched networks, using the Internet as an application. Very encouraging results have demonstrated that ant-based distributed control (stigmergetic control) is particularly suitable for a transport network layer, as far as scalability, adaptability to different network topology, and traffic patterns are concerned. This approach has also been used in the scope of the MARINER[7] project to implement a load control strategy into an intelligent network. We propose to use this approach to solve the specific problems related to OTN management.

The mobile agents paradigm refers to the ability of an agent to move from one node to another along with all the specific problems which relate to its handling (agent location,

---

[6] In the OPTIMA architecture.

[7] MARINER stands for Multi-Agent Architecture for Distributed-IN Load Control and Overload Protection – ACTS project (AC333). Web site: http://www.teltec.dcu.ie/mariner/ [06.06.2000].

security, navigation model, etc.). A lot of mobility frameworks are available on the market nowadays. Our attention has been drawn to the Grasshopper platform[8] (Breugst and Magedanz, 1998). This environment allows us to exploit the TMN-environment and thus to communicate with stationary TMN-like agents. It could be a candidate for the simulator we plan to develop in the scope of the OPTIMA project.

16.4.2 Family of optical agents

Considering both communication channels (Digital Wrapper Overhead-based, OSC Overhead-based), we have identified two families of λ-agents: channel-associated λ-agents (or λc-agents) and multiplex-associated λ-agents (or λm-agents). The choice of the family of agents will basically depend on the nature of the task to be performed. Either the task is specifically related to a wavelength, or it is specifically related to a group of wavelengths (wavelengths multiplex). λc-agents will be used for channel provisioning, while λm-agents will be used for path protection discovery, taking into account the disjoint path constraint. Full protection can actually be guaranteed only if distinct optical fibres are used between working and protection paths. Additionally, one kind of agent can be used as a backup for the other, making the transmission of agents more robust in case of failure. As far as the number of optical agents is concerned, λc-agents will probably be more numerous in the network than λm-agents, since a single fibre may encompass over 100 wavelengths in the future.

Our approach tends to define large societies of agents. The efficiency of the moving operation will consequently acquire considerable importance. That is the reason we are working on a lambda-Agent Transport Protocol (λ-ATP) which will link the overhead bytes used by our agents to the mobile agents framework. Such a protocol will avoid excessive access to the mobile agent framework and will provide multi-hop agent routing facilities.

16.4.3 Example: provisioning in OTN

The following example (Figure 16.3) describes a provisioning process in an optical network, which can typically be performed by λ-agents. It includes the path protection discovery process without wavelength allocation. The network topology is very simple but should be sufficient to illustrate the use of λ-agents. Let us consider an optical network that has two wavelengths between A-B and one wavelength on the other links. There is only one optical fibre along each node.

Here is the process description: Suppose you have as a client an SDH logical ring network along A-E-D. The client asks the TMN manager for the establishment of an A-D link with self-healing ring protection through A-E-D. This request is brought to the provisioning agency, which will have to find available wavelengths between A and D:

---

[8] Grasshopper – The Agent Platform. Web site: http://www.grasshopper.de [08.06.2000].

1. The agency initiates the path discovery process in order to find wavelength routes between A and D.

2. According to the available outputs and routing information in the local A node, new λc-agents are launched at every available output wavelength, leaving a special mark indicating that this wavelength is temporarily busy. Each agent tries to use the best-suited wavelength, according to wavelength frequency and other hardware parameters (assuming that we have two frequencies between A and B).

3. At node B, $\lambda c1$-agent is forwarded to node C (assuming that we have the same frequency between A and B and between B and C for this agent) and a new λc-agent is forwarded to node D (assuming that we have a third frequency between B and D) whereas $\lambda c2$-agent dies. All the agents will keep track of their itinerary by pushing the node address on a private stack (including the wavelength number). New agents are created by a cloning operation: a copy of the private address stack of the initiator agent is duplicated in the new agent; the latter thus inherits information about the previous travel.

4. The λc-agent that had been sent to node D has discovered its final destination. At node D, it leaves a mark informing the other agents that the route has been found; it finally travels back to the first node (A) by popping the node addresses found on its private stack. This agent is also entrusted with setting up the local MIB at each network element to establish a link between the nodes. Using the multi-hop capability of an optical agent, another high-priority cloned $\lambda c1$-agent can simultaneously start from node D with the mission of keeping the provisioning agency informed of the discovery. Thus, while the initial agent is setting up the network elements, other agents can initiate their own processes independently and as fast as possible.

5. Once the $\lambda c1$-agent is back home, the provisioning agency informs the protection agency of the protection strategy required by the SDH client. Note that the agency includes two sub-entities for fault protection and fault location. The protection agency then initiates the path protection discovery process by means of λm-agents. This agent considers only disjoint optical fibre.

6. The λm-agent performs its task of path discovering with the help of routing tables.

7. As step 6.

8. As step 6.

9. Once the λm-agent has discovered a valid path, like the λc-agent before it can immediately travel backwards to the initial node and inform the provisioning agency so that further wavelength allocation processing corresponding to the found protection path can then be performed.

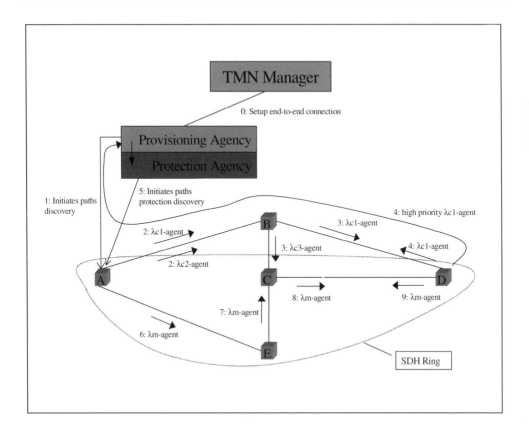

Figure 16.3 Example of provisioning with λ-agents: establishing a connection between A and D
with logical protection paths on A-E-D

16.5   Intelligent OTN Platform

This paragraph briefly describes the overall principle of the architecture and
communication schemes in the intelligent optical platform that we intend to develop in the
scope of the OPTIMA project.

In order to remain as close as possible to standardisation, which represents a major issue
in the telecommunication world, we intend to consider ongoing work currently pursued
inside the OMG (Object Management Group). MASIF[9] specifications (OMG, 1998), for
instance, should allow future large heterogeneous networks to support interoperability
between different mobile agents frameworks; as for JIDM specifications, they aim at
accessing managed objects of the local MIB in the NE with a CORBA interface.
Considering direction statements defined by the TMF, our future intelligent optical
platform should meet the expectations of the telecommunications market in terms of

---

[9] MASIF – Mobile Agent Service Interoperability Facility. Web site: http://www.fokus.gmd.de/research/cc/ecco/
masif/index.html [06.06.2000].

compliance with open interfaces and distributed systems. Figure 16.4 shows a variety of entities that are able to communicate with one another. In this example, two agencies manage a subset of λ-agents and exchange information with a manager. Three network elements (NE) can communicate with the Mobile Agent Framework (MAF). Either a proxy is used if the NE is not able to host mobile agents, or the NE provides an embedded Java Virtual Machine that facilitates the hosting of mobile agents. The communication between the proxy and the NE can be carried out with a JIDM-like gateway.

Stationary agents, such as switching-protection agents, can access the managed objects through all – or a portion of – the local MIB. According to their characteristics, stationary agents can read and write attributes and create and delete instances of managed objects. In the intelligent optical platform, stationary agents and managed objects constitute a part of the environment required for optical agents; that is, λ-agents can set up or obtain relevant information conditioning their behaviour, as far as the reactivity property is concerned (see 16.4.1).

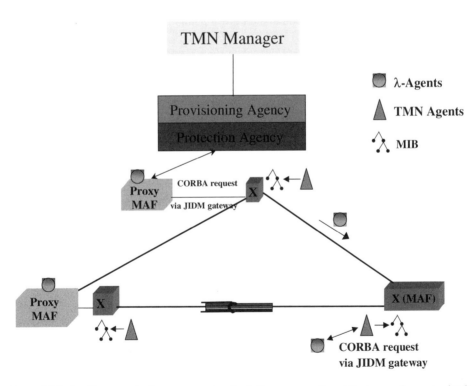

Figure 16.4 Intelligent optical transport network platform – overall architecture and communication scheme between entities

16.6   Conclusion and Future Work

In this chapter, we have described ongoing work in our Telecom Research Group, in the field of management of WDM optical networks. We are convinced that the intersection of distributed systems, artificial intelligence and future optical components, will allow us to build the future high-bit-rate optical networks in an efficient manner: future optical networks will be able to support several types of client layers with various services requirements, as well as a non-restricted logical topology, and to give robust solutions for the management of such networks. We have introduced the concept of an optical agent (λ-agent) which is able to travel efficiently within the optical network and could be prioritised using the λ-ATP protocol, thus avoiding the use of the mobile agents framework. We think that putting optical agent societies within the network and letting them perform simple tasks will allow us to solve complex problems such as fault protection, fault location and provisioning with multi-constraint requirements. We also plan to define a new class of optical agents in order to solve OTN performance management issues. Convergence time is still an open issue, but results of experiments conducted by other research institutes have showed that routing algorithms based on agent societies are very encouraging. The approach presented above will be developed and experimented in detail in the scope of a proposed Swiss National Science Foundation project called OPTIMA. In the same context, we intend to develop a simulator of multi-layer OTN architecture (a separate SDH and IP network on top of a common OTN). The Grasshopper platform, which supports communication between mobile agents and stationary TMN-like agents, could serve as a basis for our simulator. Optical agents will then be defined and injected into the platform to perform the complex tasks of fault protection and location. We also intend to solve open issues like the choice between λc-agents and λm-agents, according to the nature of the problem.

10.7   References

Bieszczad, A., Pagurek, B. and White, T. (1998) Mobile Agents for Network Management. *IEEE Communications Surveys*, September issue.

Breugst, M. and Magedanz, T. (1998) On the Usage of Standard Mobile Agent Platforms in Telecommunication Environments, in *Proceedings of Intelligence in Service & Networks (IS&N '98)*, Antwerp, Belgium.

Crochat, O. (1998) Wavelength Division Multiplexing Networks and Failure Protection. PhD thesis Number 1851, Ecole Polytechnique Fédérale de Lausanne.

Di Caro, G. and Dorigo, M. (1998) AntNet: Distributed Stigmergetic Control for Communications Networks. *Journal of Artificial Intelligence Research*, **9**, 317–365.

Hayzelden, A.L.G. and Bigham, J. (1999) Agent Technology in Communications Systems: An Overview. *The Knowledge Engineering Review*, **14(3)**, 1–35.

Huitema, C. (1998) *IPv6 – The New Internet Protocol* (2nd Edition). Prentice-Hall.

ITU (1996) Recommendation M.3010 (05/96) – Principles for a Telecommunications management network. Available: http://www.itu.int/itudoc/itu-t/rec/m/m3010.html [06.06.2000].

ITU (1998) Recommendation G.841 (10/98) - Types and characteristics of SDH network protection architectures. Available: http://www.itu.int/itudoc/itu-t/rec/g/g800up/g841.html [06.06.2000].

ITU (1999) Recommendation G.872 (02/99) – Architecture of optical transport networks. Available: http://www.itu.int/itudoc/itu-t/rec/g/g800up/g872.html [06.06.2000].

ITU (2000) Recommendation G.805 (03/00) – Generic functional architecture of transport networks. Available: http://www.itu.int/itudoc/itu-t/rec/g/g800up/g805.html [06.06.2000].

Lagasse, P., Demeester, P., Ackaert, A., Van Parys, W., Van Caenegem, B., O'Mahony, M., Stubkjaer, K. and Benoit, J. (1998) A European View by the HORIZON Project and the ACTS Photonic Domain. Available: http://www.intec.rug.ac.be/horizon/projects/horizon.html [06.06.2000].

Lucent Technologies (1999) A Proposal for Providing Channel-Associated Optical Channel Overhead in the OTN. Available: http://www.t1.org/index/0816.htm [06.06.2000].

Mas, C. and Le Boudec, J.-Y. (1997) An Alarm Filtering Algorithm for Optical Communication Networks, in *Proceedings of MNS97*, pp. 205–218, Montreal.

OMG (1998) Mobile Agent System Interoperability Facilities Specification (Joint Submission). Available: http://www.fokus.gmd.de/research/cc/ecco/masif/documents.html [08.06.2000].

Sexton, M. and Reid, A. (1997) *Broadband Networking: ATM, SDH and SONET*. Artech House, Boston.

Wooldridge, M. and Jennings, N.R. (1995) Agent Theories, Architectures, and Languages: A Survey, in M. Wooldridge and N.R. Jennings (Eds.) *Intelligent Agents* (Lecture Notes in Computer Science 890), pp. 1–39. Springer-Verlag.

Yoo, J.Y. (1997) Design, Analysis, and Implementation of Wavelength-Routed All-Optical Network Routing and Wavelength Assignment Approach, AT&T Laboratories, Holmdel, NJ. Available: http://www.comsoc.org/pubs/surveys/Yoo/yoo.html [08.06.2000].

17

How to Move Mobile Agents

P. Alimonti, F. Lucidi, S. Trigila

17.1   Introduction

Following the competition and convergence between traditional telecommunications systems and the Internet, network systems are becoming more and more sophisticated and complex. In order to cope with this complexity, distributed object technology has been widely used and the client/server paradigm has been recognised as the key model for service design and deployment. Servers perform the needed processing and then pass back operation results to the requesting clients. This approach, however, may have some drawbacks in terms of efficiency and cost, especially when the amount of data exchanged and the number of interactions are high and/or the bandwidth available from the underlying transport infrastructure is limited. Agent technology is considered an extension of the distributed object technology towards overcoming the limitations of the server/client paradigm (Jennings and Wooldridge, 1998).

An *agent* is a program delegated to perform a given task for a client, a human user or another program, which may be located in a network node distinct, and possibly far apart, from the node where the client resides. An agent can be endowed with a certain degree of intelligence that allows it to take actions autonomously. An *intelligent agent* is able to decide by itself how to proceed in response to its environment, only directed by its own goals and knowledge, thus avoiding asking the user for instructions. This reduces considerably the number of interactions in the network.

Properties normally quoted for agents are: independence of action, planning ahead, ability to respond to changing circumstances, flexible interaction with other agents, ability to learn, ability to negotiate. Main benefits expected from the use of agents are: provision of natural abstraction to model relevant problems in several domains, unifying technology (combining advantages from object orientation, distributed computing, parallel processing, symbolic processing, and artificial intelligence), and powerful structuring of problem knowledge representation and behaviour code.

An agent can also be provided with the ability to travel in the network, i.e. the possibility to move from one node to another during the task's execution. In this case we speak of a *mobile agent*. A mobile agent migrates to another node according to what is scheduled to accomplish its task. It should be noted that intelligence and mobility of agents are orthogonal properties: an agent may be mobile without being intelligent, and vice versa.

For a background on properties and advantages of mobile agents, we refer the reader to outstanding works as Lange and Oshima (1998) and Brewington et al. (1999). We take the opportunity to quote a passage of the latter work: 'Most distributed applications fit naturally into the mobile-agent model, since a mobile agent can migrate sequentially through a set of machines, send out a wave of child agents to visit machines in parallel, remain stationary and interact with resources remotely or any combination of these three extremes. Complex, efficient and robust behaviours can be realised with surprisingly little code'.

Foreseen applications of mobile agents are quite attractive, particularly in the telecommunications field, with the aim of better exploitation of network resources or better-quality service assurance, compared with what is available nowadays (Eurescom Project, 1999). Possible applications in the telecommunications area include, but are not limited to, the following: proxying and tracking of mobile terminals and users, network management (e.g. fault monitoring and alerting), service management (e.g. negotiation of parameters on behalf of the user or the provider), information retrieval (e.g. query-based filtering at the server site, and minimisation of data returned to a client), product selection in an electronic commerce environment (e.g. evaluation of best offers, following visits to a relevant number of sites).

In some of these applications (for instance, electronic commerce) it could be necessary to visit a considerable number of sites. A straightforward solution would consist in sending as many agents in parallel as there are sites to visit. However, when the number of sites is high, this solution may be too costly, in terms of network traffic, and not worth the relatively short response time. An efficient solution is to have a number of agents each touring through a selected subset of sites. We then can consider a particular subset of sites as a network on its own, and speak of an agent visiting them as a complete tour. While visiting a site, the agent performs a fixed and, possibly, an extra task.

In this scenario, the benefit introduced by the use of mobile agents strictly depends on the planning of the agent tours, that is on choosing the sequence of the sites to be visited by each agent that minimises the cost of the overall task. We assume that a mobile agent is able to implement a tour scheduled by itself or by another program it can communicate with, prior to starting its tour.

Considering time as a measure of cost, we investigate two classes of tour-planning problems. In the first class, the target is to minimise the time for touring the network, i.e. the completion time for visiting all sites. In the second class, the aim is to minimise the latency of the network, i.e. the sum of the waiting times of every site where an extra task is needed, before completing the work at that site.

The chapter is organised as follows. In section 17.2, we establish some basic definitions and formulate general objective functions for the two problems of network tour and network latency. In section 17.3 we discuss feasibility and complexity of optimisation techniques for the two problems, and introduce significant sub-cases where efficient solutions can be found. Finally, in section 17.4 we summarise the results presented in the chapter.

17.2   Defining Tours for Mobile Agents

An early approach to mobile agent planning was mainly directed at finding heuristic solutions by means of artificial intelligence techniques Ambros-Ingerson and Steel (1972), Drummond and Bresian (1990). More recently, in Moizumi and Cybenko (2000), a specific network tour-planning problem that arises from information retrieval applications has been tackled making use of combinatorial optimisations techniques and of probabilities of finding the desired information at the visited nodes. Our chapter is based on a similar approach, but defines probabilities in a more general way and addresses also the problem of network latency.

17.2.1   Basic definitions

As a starting point, we define a planning problem, which we shall call a *Mobile Agent problem*, in the following terms. Let S be a set of sites in a network and let us consider a mobile agent that moves from one site to another, starting from a site s_0. Travel between a pair of sites takes some time. Let $t_{i,j}$ be the time duration, assumed constant, needed to move from site s_i to s_j. At each s_i the agent has to perform some actions that take a time c_i (compute time) and, with probability p_i, has to carry out extra processing that requires a time e_i (extra compute time). The probability p_i may change during agent travel depending on the sites visited before s_i. A tour starting from site s_0 is a sequence of sites $\pi = \{s_{i_0}, s_{i_1}, s_{i_2}, ..., s_{i_n}\}$, where $[i_0, i_1, i_2, ..., i_n]$ is a permutation such that $s_{i_0} = s_0$. A feasible solution of the Mobile Agent problem is a tour starting in s_0 and visiting all sites in S. An optimal solution of the Mobile Agent problem is a tour that minimises the cost of the agent task. In sections 17.2.2 and 17.2.3 we define two objective functions expressing two different types of cost: (a) time for touring the network, and (b) network latency.

The rationale for assuming variation of probabilities is the following. Sites may influence each other for several reasons. As an academic example, consider an agent whose task is to check if a set of sites is infected by viruses and clean them. If a virus is found on a site it is reasonable that the probability of finding the same virus in another 'near and/or similar' site, and consequently the probability of doing extra clean-up work, increases. A more realistic example can be found in information retrieval applications, where a mobile agent is required to accomplish a certain task that includes the search for information. The fact that a piece of information is found on a site, and therefore elaborated, may decrease the probability of performing extra work in the next sites to be visited.

17.2.2   Touring the network

In this type of problem the goal of the agent is to visit a given set of sites in a network and eventually return to the departure site, minimising the overall completion time. The problem is typically relevant in applications where an agent visits the sites in order to look up, elaborate and package information on a certain subject or certain goods. The tour is

justified by the circumstance of information being distributed over various sites, with more
than one site able to offer the same piece of information or the same goods.

Consider an agent whose goal is retrieving information from the simple network of
three machines as shown in Figure 17.1.

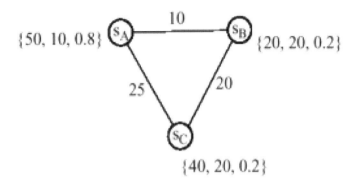

Figure 17.1 A simple example of a network

Arc labels represent the traversal times between machines. Nodes are labeled by a triple
which expresses the look-up time, the elaboration time and the initial probability of finding
a useful piece of information, respectively. For the sake of simplicity, the starting site is not
drawn and the traversal time from it to every other site is assumed to be null. Besides, we
suppose that if some information is found at site s_B the probability of finding information
at site s_A is halved. Thus, the next probability function is

$$up_p(s, P) = \begin{cases} P & \text{for} \quad s = s_A, s_C \\ P' & \text{for} \quad s = s_B \quad \text{such that} \\ & P'[s_A] = P[s_A](1 - 0.5 \cdot P'[s_B]) \quad P'[s_B] = P[s_B] \quad P'[s_C] = P[s_C] \end{cases}$$

It is noteworthy that if s_A precedes s_B in the tour then the probability of finding
information at s_A when s_A is visited is still unchanged with respect to the initial one. On
the other way round, if s_B is visited before s_A the probability of s_A will decrease from 0.8
to 0.72. In fact, after visiting s_B the probability of s_A becomes 0.4 with probability
$P[s_B] = 0.2$, and remains unchanged with probability $1 - P[s_B] = 0.8$. Assuming that the
agent migrates from machine to machine in the order of s_A, s_B and s_C the expected time
to complete the tour is:

$$\Gamma_{s_A,s_B,s_C} = (\text{look-up time for } s_A + \text{Pr. for info in } s_A \times \text{collection time for } s_A)$$

$$+ (\text{look-up time for } s_B + \text{Pr. for info in } s_B \times \text{collection time for } s_B + \text{time from } s_A \text{ to } s_B)$$

$$+ (\text{look-up time for } s_C + \text{Pr. for info in } s_C \times \text{collection time for } s_C + \text{time from } s_B \text{ to } s_C)$$

$$= (50+0.8 \times 10) + (20+0.2 \times 20+10) + (40+0.2 \times 20+20) = 156.$$

It is easy to see that, changing the visit order a different result may be obtained. In particular, for the tour $\{ s_0, s_C, s_B, s_A \}$ the time needed will decrease, since the probability of finding information at s_A is 0.72 when it is visited in this tour.

We shall call Minimum Travelling Agent problem (MIN-TAP in the remainder of this chapter) the problem of deciding a tour π whose expected time for completion is minimum. Consider an instance of MIN-TAP and a tour $\pi = \{s_0, s_{i_1}, s_{i_2}, ..., s_{i_n}\}$. The average time needed to complete a tour π is the sum of all travel times from site to site, all compute and all average extra compute times of that tour. However, since the sum of the compute times is constant for any tour and the goal of the problem is to minimise the variable part of the cost of tour, we have

$$C_\pi = \sum_{k=1}^{n} (t_{i_{k-1},i_k} + p_{i_k|i_1,i_2,...,i_{k-1}} \times e_{i_k})$$

where $p_{i_k|i_1,i_2,...,i_{k-1}}$ is a conditional probability, which expresses the probability of extra compute time associated to s_{i_k} given that the agent has already visited the first $k-1$ sites in the tour π. The above function is too complex to evaluate in general. It is the subject of an on-going study by the authors. An interesting case where the computation of this function can be simplified is when $p_{i_k|i_1,i_2,...,i_{k-1}} = p_{i_k|i_{k-1}}$. This assumption is similar to the one underlying the state transition probabilities in a Markov chain. The cost of a tour $\pi = \{s_0, s_{i_1}, s_{i_2}, ..., s_{i_n}\}$ then becomes:

$$C_\pi = \sum_{k=1}^{n} (t_{i_{k-1},i_k} + p_{i_k|i_{k-1}} \times e_{i_k}) \tag{17.1}$$

For the sake of concreteness we give an example of how the $p_{i_k|i_{k-1}}$ can be computed. Without loss of generality let us use the permutation $[1, 2, ..., n]$. Let α_{ij} be a 'neighbouring factor' from site s_i to site s_j, representing the influence of extra computations at site s_i on the need of extra computations at site s_j. Again without loss of generality we can say that if extra computation occurs at site s_i, then the probability of extra computation needed at site s_j changes by a factor α_{ij}, otherwise the latter probability does not change. The probability of extra computation needed at site s_j after visiting s_i can be therefore expressed as:

$$p_{j|i} = \alpha_{ij} \times p_{j|i-1} \times p_{i|i-1} + p_{j|i-1} \times (1 - p_{i|i-1}) \qquad (17.2)$$

In this example, all that is needed to start the process is an initial set of probabilities of extra computation associated to all sites when the agent is at s_0 and a neighbouring matrix

.

17.2.3   Latency of the network

This class of problems models important applications such as the management of failures in a telecommunication network whose machines can be affected by hardware and software exceptions. In particular, we may consider a mobile agent that travels in such a network for checking and restoring out-of-order machines. In this context it is often strategic to minimise the sum of the average time that those machines have to wait before being repaired by the agent (expected latency), rather than the length of the agent tour.

Consider now an agent whose task is the management of failures in the network shown in Figure 17.1. Here, the values associated to each node represent the check time, the repair time and the probability of corruption, respectively. We also suppose that the corruption of one machine does not influence at all the probabilities of corruption of the others, i.e. the next probability function is the identity. Assuming that the agent migrates from machine to machine in the order of s_A, s_A, and s_A the times W that each corrupted machine has to wait for being restored are:

$W_{s_A} =$ time to check s_A + time to repair s_A

$W_{s_B} =$ time to check s_A + Pr. of s_A corruption × time to repair s_A + time to go

         from s_A to s_B + time to check s_B + time to repair s_B

$W_{s_C} =$ time to check s_A + Pr. of s_A corruption × time to repair s_A + time to go

         from s_A to s_B + time to check s_B + Pr. of s_B corruption × time to repair s_B +

         time to go from s_B to s_C + time to check s_C + time to repair s_C

and the total expected latency of this tour is

$L_{s_A, s_B, s_C} =$ Pr. of s_A corruption × W_{s_A} + Pr. of s_B corruption × W_{s_B} + Pr. of s_C

corruption × $W_{s_C} = 0.8 \times 60 + 0.2 \times 98 + 0.2 \times 172 = 102$

It is easy too see that changing the visit order yields a different result.

We shall call Minimum Latency problem (MIN-LAT) the problem of deciding a tour π that minimises the sum of the average times that sites have to wait until extra computation is carried out.

Let W_{π,i_k} be the average time a site s_{i_k} has to wait until extra computation is carried out on it. W_{π,i_k} is made up of the following terms:

1. time to travel from s_0 to s_{i_k};

2. time to check all the visited sites including s_{i_k};

3. average extra compute time at all sites visited before reaching s_{i_k};

4. extra compute time at s_{i_k}.

Therefore:

$$W_{\pi,i_k} = \sum_{h=1}^{k} t_{i_{h-1},i_h} + \sum_{h=1}^{k} c_{i_h} + \sum_{h=1}^{k-1} p_{i_h|i_{h-1}} \times e_{i_h} + e_{ik}$$

where, once again, $p_{i_h|i_{h-1}}$ is the probability of extra compute time of site s_{i_k} when it is visited in the tour π. The expected latency L_π of the tour is the sum of the W_{π,i_k} for all sites weighted with the probabilities of extra compute time.

$$L_\pi = \sum_{k=1}^{n} p_{i_k|i_{k-1}} \times W_{\pi,i_k} \qquad (17.3)$$

17.3   Optimising Tours for Mobile Agents

This section is devoted to an evaluation of feasibility and complexity of possible solutions to the problems introduced in the previous section.

17.3.1   Minimising network tour cost

MIN-TAP generalises a classical optimisation problem, known as the Minimum Travelling Salesperson (MIN-TSP) problem (Garey and Johnson, 1979). Indeed, if the values of the probability associated to the sites do not change while the agent is visiting the network, MIN-TAP reduces to MIN-TSP. With constant probabilities, to minimise the objective described in equation (1) is equivalent to minimising:

$$C_\pi = \sum_{k=1}^{n} t_{i_{k-1},i_k}$$

MIN-TSP, has been shown to be NP-hard, i.e. solvable in general only with algorithms requiring non-polynomial computational resources. Therefore, MIN-TAP is also NP-hard.

In this work, we limit our analysis to a subclass of MIN-TAP defined by assuming traversal times all equal to a constant or negligible with respect to compute times. Compared to MIN-TSP where the term depending on probabilities does not contribute to the objective,

objective, here the traversal times do not impact the objective. Therefore, to minimise the cost of the tour is equivalent to minimising:

$$C_\pi = \sum_{k=1}^{n} p_{i_k|i_{k-1}} \times e_{i_k}$$

Assuming $\pi = \{s_0, s_1, s_2, ..., s_n\}$ and expressing the terms $p_{i_k|i_{k-1}}$ as in (17.2), we can also write

$$p_{j|i} = (\alpha_{ij} - 1) \times p_{j|i-1} \times p_{i|i-1} + p_{j|i-1} = \beta_{ij} \times p_{j|i-1} \times p_{i|i-1} + p_{j|i-1}$$

where $= -1$ describes a graph that we shall call the influence graph. We now consider significant cases where the influence graph has particular properties.

17.3.1.1 The influence graph is a star

To discuss this case, let us start by assuming that there is only site s_h influencing another site s_k, i.e. $\beta_{hk} \neq 0$ and $\beta_{ij} = 0$ for $i \neq h$ and $j \neq k$. It can be shown that the cost of the tour only depends on the order between site s_h and site s_k in the tour, regardless of the position of other sites. In particular if $\beta_{hk} < 0$ the probability associated to s_k decreases when visiting s_h and then s_k must follow s_h. Conversely, if $\beta_{hk} > 0$ the above probability increases and then s_k must precede s_h. The order of visiting the other sites does not impact the cost of the tour.

We can easily extend the reasoning to the case in which s_h influences more than any another sites. It can been shown that an optimal tour is a partial order between s_h and each site influenced by s_h. For each s_j such that $\beta_{hj} < 0$, s_j must follow s_h, conversely for each s_j such that $\beta_{hj} > 0$, s_j must precede s_h. The position of the other sites in the tour does not impact the cost.

17.3.1.2 The influence graph is a directed acyclic graph (DAG)

A further generalisation that includes the previous case is when the relationships between sites have a hierarchical structure. In other words, the inter-dependence between the probabilities associated to the sites fulfils the following property. If the visit of one site s_i affects the probability of another site s_j, then s_j cannot affect the probability of s_i, either directly or indirectly.

In this case it is possible to efficiently determine tours for which the agent moves in such a way as to avoid probability changes in one site before visiting it.

In order to construct a tour π that satisfies such a property we follow the partial order displayed by the influence graph. Since this graph is a DAG there must be at least one node without arcs leaving it. Then we choose the site corresponding to one of these nodes as the first to visit in π. We remove that node and all input arcs to it. Of course, the remaining graph is still a DAG. Therefore, we iterate the previous procedure in order to determine the second site to visit and so on. This algorithm can be implemented with time complexity that is linear in the size of the problem.

Observe that, in general, the tour determined is not unique, since there can be more than one tour satisfying the partial order displayed by the influence graph. Nevertheless, the cost of all these tours, for which the agent moves so as to avoid probability changes in one site before visiting it, is the same. It is given by

$$\sum_{i=1}^{n} p_i \times e_i$$

which we refer to as the basic cost of the problem. An interesting special subcase of MIN-TAP with a directed acyclic influence graph is when all the probability changes are positive, i.e. $\beta_{ij} \geq 0$. In this case, any tour of basic cost is an optimal tour, as well. Unfortunately, such a result does not extend to networks where values of the influence factor can arbitrarily change. This is because the presence of negative influence factors gives rise to probability decreases and consequently to a reduction of compute times, but the global amount of these reductions depends on the order in which nodes are visited. Nevertheless, we believe that tours of basic cost deserve attention in mobile agent problems, even if they represent suboptimal solutions. In fact, the basic cost can be used as a reference figure to assess whether and how much a further planning of the agent route is worth the achievable reduction. Furthermore, we guess that a tour of basic cost obtained by a simple and low-cost algorithm is suitable as a good starting point by heuristics for finding optimal or nearly optimal solutions.

17.3.1.3   The influence graph is a directed graph with elementary cycles

In this case we allow the presence of cycles of length 2. The simplest subcase is when we have two sites s_h and s_k, such that $\beta_{hk} \neq 0$ and $\beta_{kh} \neq 0$, while all the other influence factors equal 0. We can show that if $\dfrac{e_h}{\beta_{hk}} < \dfrac{e_k}{\beta_{kh}}$ a tour is optimal if s_k follows s_h, and conversely. The ordering of the other sites does not impact the cost. A straightforward generalisation is when there are more pairs of sites that influence each other, but no pair intersects. For each pair of sites we act as above, giving rise to a total order inside any pair and to a partial order among all the sites in the network. Then, every tour that fulfils such a partial order is optimal. We conclude by observing that the last result can also be extended to the case in which we have, not pairs, but n-tuples of sites that mutually influence each other (i.e. any site belongs to at most $n-1$ pairs). However, for large n the cost of finding optimal total orders inside the n-tuples can become extremely hard.

17.3.2   Minimising network latency

The MIN-LAT generalises an optimisation problem known as the Minimum Travelling Repairperson problem (MIN-TRP) (Afrati et al. 1986; Koutsoupias, Papadimitriou, and

Yannakakis, 1996). In particular, MIN-LAT becomes MIN-TRP when all probabilities are constant and equal to 1:

$$L_\pi = \sum_{k=1}^{n} \sum_{h=1}^{k} (t_{i_{h-1},i_h} + c_{i_h} + e_{i_h})$$

TRP is known to be NP-hard.[10] Of course, MIN-LAT cannot be easier. Nevertheless, interesting results can be achieved under *the particular assumption of constant probabilities of extra compute times*. This assumption models, for example, a network situation where the influence between sites is extremely loose, and therefore the probability values associated to the sites remains almost constant during the tour. Also, whenever the agent lacks knowledge about site inter-influence, it may be useful to attack the problem by considering that probabilities are constant. Within this case we consider two further subcases: (a) all traversal times equal to a constant; (b) all (extra) compute times negligible compared to traversal times.

17.3.2.1 Latency with constant probabilities and equal traversal times
Under this assumption equation (3) becomes

$$L_\pi = \sum_{k=1}^{n} p_{i_k} \cdot \left(kt + \sum_{h=1}^{k} c_{i_h} + \sum_{h=1}^{k-1} p_{i_h} \cdot e_{i_h} + e_{i_k} \right)$$

We can show that the optimal solution can be calculated in polynomial time and corresponds to a tour by which the agent visits the sites in increasing order of the ratios

$$\frac{t + c_{i_k} + p_{i_k} \times e_{i_k}}{p_{i_k}} \qquad (p_{i_k} \neq 0).$$

Finally, we observe that when the traversal times and the fixed compute times are insignificant with respect to the extra compute times an optimal tour displays the sites sorted in increasing order of extra compute time.

17.3.2.2 Latency with constant probabilities and negligible compute times
In this case the latency can be expressed as:

$$L_\pi = \sum_{k=1}^{n} p_{i_k} \cdot \sum_{h=1}^{k} t_{i_{h-1},i_h}$$

---

[10] Intuitively speaking, it seems more complex than MIN-TAP due to the non-local nature of the objective function: a raise in the cost of visiting a site at the beginning of the tour affects the latency of all the remaining sites.

Unfortunately, this is a weighted variant of MIN-TRP. Therefore, the assumption of negligible compute times does not bring any simplification and the problem still remains NP-hard.

17.4   Conclusions

The main contribution of this chapter is the introduction of a computational framework for representing and solving mobile agent planning problems occurring in particular applications. Specifically, we have focussed our attention on networks where a mobile agent has to visit a given set of sites in order to accomplish a certain task. While visiting a site, the agent performs a fixed amount of work and with a certain probability extra work.

In particular, we have defined the Minimum Travel Agent (MIN-TAP) and Minimum Latency (MIN-LAT) problems. In the former, the goal of the agent is to minimise the time for touring the network, that is, the completion time for visiting all sites. In the latter, the aim is to minimise the latency of the network, that is, the sum of the waiting times of every site where an extra task is needed, before completing the work at that site. To reflect real-world situations, the probability of performing extra work when visiting a site has been assumed to be changing while the agent is travelling.

MIN-TAP and MIN-LAT have been shown to be non-trivial generalisations of two well-known problems, the Minimum Travelling Salesperson problem (MIN-TSP) and the Minimum Travelling Repairperson problem (MIN-TRP), respectively. Therefore, known optimal algorithms for our two problems require non-polynomial computational resources. However, for subclasses of problems of practical significance - yet really distinct from MIN-TSP and MIN-TSP we have shown that optimal solutions can be found by means of polynomial algorithms. Those subclasses were obtained by making suitable assumptions on the so-called influence factor, on the behaviour of probabilities and on the compute times.

The proposed framework is an initial milestone in an on-going research path that aims to find new conditions of applicability of optimisation methods to the above problems and applying the related techniques to real case studies.

Acknowledgement

Work was partially carried out under an agreement between Fondazione Ugo Bordoni and ISCTI (Istituto Superiore per le Comunicazioni e le Tecnologie dell'Informazione), Rome, Italy.

17.5   References

Afrati, F., Cosmadakis, S., Papadimitriou, C., Papageorgiou, G., and Papakostantinou, N. (1986) The complexity of the traveling repairman problem, *ITA* **20(1)**, 79-87.

Ambros-Ingerson, J. and Steel, S. (1972) Integrating planning, execution and monitoring, in *Proceedings of the Seventh National Conference on Artificial Intelligence* (AAAI-88), Morgan Kaufmann, Saint Paul, Minnesota.

Brewington, B. and Gray, R. (1999) Mobile Agents in Distributed Information Retrieval, in *Intelligent Information Agents* (M. Klusch, ed.), Springer, pp. 355-395.

Drummond, M., and Bresian, J.: (1990) Anytime synthetic projection: maximising the probability of goal satisfaction, in *Proceedings of AAAI-90*.

Eurescom Project P712 (1999) The Applicability of Intelligent and Mobile Agents to Service and Network Management, Deliverable 1, Vol.1 (Main Part), Heidelberg (Germany), June.

Garey, J. and Johnson, D. S. (1979) *Computers and Intractability: a Guide to the Theory of NP-Completeness*, Freeman & Co., San Francisco.

Jennings, N.R. Wooldridge, M.J. (eds.) (1998) *Agent Technology - Foundations, Applications and Markets*, Springer, Heidelberg (Germany).

Koutsoupias, E., Papadimitriou, C., and Yannakakis, M. (1996) Searching a fixed graph, in *Proceedings of the 23rd Int. Colloquium on Automata, Languages and Programming* (ICALP-96), vol. **1099**, Lecture Notes in Computer Science, pp. 280-289.

Lange, D.B, and Oshima, M. (1998) *Programming and Deploying Java Mobile Agents with Aglets*, Addison-Wesley.

Moizumi, K. and Cybenko, G.: (2000) The travelling agent problem, to appear in *Mathematics of Control, Signals and Systems*.

18

Market-Based Call Routing in Telecommunications Networks Using Adaptive Pricing and Real Bidding

M. A. Gibney, N. R. Jennings, N. J. Vriend, J. M. Griffiths

18.1   Introduction

In telecommunications networks, call traffic is typically routed through the network from source to destination on the basis of information about the traffic on that path only. Therefore, path routing is carried out without regard to the wider impact of local choices. The main consequence of this myopic behaviour is that under heavy traffic conditions the network is utilised inefficiently – rejecting more traffic than would be necessary if the load were more evenly balanced. One means of performing such load balancing is to centrally compute optimal allocations of traffic across the network's paths using predictions of expected traffic (Bertsekas and Gallager, 1987). When such calculations have been completed, the network management function can configure the network's routing plan to make the best use of the available resources given the predicted traffic. However, as networks grow larger and involve more complex elements, the amount of operational data that must be monitored and processed (by the network management function) increases dramatically. Therefore in centralised architectures, management scalability is bounded by the rate at which this data can be processed (Goldszmidt and Yemini, 1998). In addition, there are a number of well-known shortcomings with algorithms to compute optimal network flows; these include progressively poorer performance in heavily loaded networks and unpredictable oscillation between solutions (Kershenbaum, 1993). Furthermore, the very centralisation of the network management function provides a single point of failure, thus making the system inherently less robust.

For the above mentioned reasons, a decentralised approach to routing is highly desirable. In such cases, decisions based on more localised information are taken at multiple points in the system. The downside of this, however, is that the local decisions have non-local effects. Thus, decisions at one point in the system affect subsequent decisions elsewhere in the system. Ideally, localised control would take place in the presence of complete information about the state of the entire system. Such a state of affairs would enable a localised controller to know the consequences of a choice for the rest of the network. However, there are two main reasons why this cannot be realised in practice. Firstly, the network is dynamic and there is a delay in propagating information. This means that a model of the network state held at any one point is prone to error and difficult to keep up-to-date. Secondly, the scaling issues involved in making flow optimisation computations for the entire network (noted above) would obtain here also. Therefore, a system in which local decision-making takes place in the presence of an incomplete view of the wider network is the only feasible solution for providing distributed control.

A promising approach that combines the notion of local decision making with concerns for the wider system context is that of agent technology (Jennings and Wooldridge, 1998). Agents address the scaling problem by computing solutions locally, based on limited information about isolated parts of the system, and then using this information in a social way. Such locality enables agents to respond rapidly to changes in the network state, while their sociality can potentially enable the wider impact of their actions to be coordinated to achieve some socially desired effect. Systems designed to exploit the social interactions of groups of agents are called multi-agent systems (MAS). In such systems, each individual agent is able to carry out its tasks through interaction with a small number of acquaintances. Thus, information about the extent of the system is distributed along with whatever functionality the MAS is designed to perform.

One agent-based technique that is becoming increasingly popular as a means of tackling distributed resource allocation tasks is market-based control (Clearwater, 1996). In such systems, the producers and consumers of the resources of a distributed system are modelled as the self-interested decision-makers described in standard microeconomic theory (Varian, 1992). The individual agents in such an economic model decide upon their demand and supply of resources, and on the basis of this the market is supposed to generate an equilibrium distribution of resources that maximises social welfare. In market-based control, the metaphor of a market economy as a system computing the behaviour that solves a resource allocation problem is taken literally and distributed computation is implemented as a market price system. That is to say, the agents of the system interact by offering to buy or sell commodities at given prices (Wellman, 1996). In our case, such an approach has the advantage that ownership and accountability of resource utilisation are built into the design philosophy. Thus, market-based solutions can be applied to the management of multi-enterprise systems without forcing the sub-system owners to cooperate on matters affecting their own commercial interest.

Within this context, this chapter describes a system to balance traffic flow through the paths of a logical network, based on the local action of agent controllers coupled with their social interaction as modelled by a computational market. The approach presented here builds upon the preliminary work reported in Gibney and Jennings (1998) in that it shares the same architecture, roles and deployment model for the agents. However, to improve upon our earlier results we devised a new approach to the way that agents adapt their pricing and inventory strategies according to the outcome of individual market actions and the profitability of trading in the market. More generally speaking, this research extends the state of the art in market-based control in the following ways. First, it models a complex two-level economy, in which not only end users but also the internal components of the system compete with one another for resources. The rationale behind using a two-level economic model is to realise call admission control in the same framework as the network management function. This is novel, as market-based control has not previously been used to address two control issues in the same system. Particularly, having two kinds of market within the economy, with agents active in different roles in each of them, provides an elegant way to acquaint agents with one another. This architecture also provides an appropriate way to situate the intelligence of the system in a multi-enterprise network with self-interested enterprises. Secondly, a novel approach is adopted to pricing strategy. Our agents adapt to the outcomes of market interactions which use real bids and offers (i.e. agents state a price in an auction-like market and are then committed to buying or selling at that price in that session). This approach was adopted because it eliminates the lengthy series of interactions between agents that is required to calculate the equilibrium price in the market. Rather, we use real bidding and allow the agents to adapt their bidding behaviour to the outcomes of the auctions over time. Real bidding allows us to use more rapid (one-shot) auction protocols as markets.

The remainder of this chapter is structured in the following manner. We discuss the background and motivation for this work in section 18.2. Section 18.3 describes the architecture of the system as a whole and the institutional forms of the possible interactions between agents. The design of individual agents is given in section 18.4. Section 18.5 discusses the experiments carried out to evaluate the performance of the system. Finally, section 18.6 details our conclusions on the work presented and discusses the open issues and future work.

18.2   Background and Motivation

Decentralised approaches to routing, usually in packet-switched telecommunications networks, based on the interaction of controllers distributed through the network, have existed for some time (Schwartz and Stern, 1989). However, agent-based approaches extend this idea by modelling the interaction of distributed controllers as a social process. A number of agent-based solutions have been proposed to the problem of load balancing in telecommunications networks. Appleby and Steward (1994) make use of mobile agents roaming the network and updating routing tables to inhibit or activate routing behaviours. Schoonderwoerd, Holland and Bruten (1996) extend and improve this approach by using ant-like mobile agents that deposit 'pheromones' on routing tables to promote efficient routes (the majority of ants use the efficient routes and the pheromones re-enforce this behaviour in other ants). Hayzelden and Bigham (1998) employ a combination of reactive and planning agents in a heterogeneous architecture to reconfigure route topology and capacity assignments in ATM networks. All of these systems exhibit increased robustness and good scaling properties compared to centralised solutions. Indeed, in network environments with symmetric traffic requirements, ant-like agent solutions have even been shown to provide superior load balancing to both statically routed networks and a more conventional mobile agent approach (Schoonderwoerd et al., 1996).

However, all the aforementioned approaches model networks as a single resource and therefore act to optimise utilisation of that resource. This makes sense because a poorly managed telecommunications network benefits no one. However, the main disadvantage of such a perspective arises when different telecommunications operators join their networks together (something which is an increasingly common trend). In such cases, if the different sub-network owners agree on a single, unified static network management policy, it is unlikely that this policy will benefit all their interests individually as well as collectively over time. We address this issue by modelling our agents as the resources and groups of resources that enterprises might own or lease in a multi-enterprise environment.

Another increasingly common aspect of modern telecommunications deployment is the practice of enterprises in other sectors (banking and other traditional consumers of telecommunications services) leasing bandwidth from telecommunications providers to create virtual private networks in the short, medium and long terms (Cisco, 1999). Again, this promotes the creation of multi-enterprise networks. In such environments each enterprise clearly has an incentive both to see that over use does not degrade network performance and to make the greatest possible use of their network ownership. Since these parties cannot agree each traffic policy decision individually, conflicting incentives must be reconciled outside the traffic management domain. Typically this is achieved by allowing sub-network owners to set policy within the remit of their own resources (Stallings, 1997). However, the static nature of these policies and the conflict between them at sub-network interfaces often causes institutionalised under-use of the network as a whole.

Both of these trends suggest that telecommunications network management, once a centralised and monolithic undertaking, will increasingly benefit from an open, robust, scalable and inherently multi-enterprise approach. Therefore, one of the aims of this work is to use the multi-agent system paradigm to address the problem of multi-enterprise ownership of the network, while simultaneously addressing the problems of robustness and scalability. Against this background, the resource allocation problem in a network with multiple, non-cooperating enterprises can be recast as the problem of reconciling competition between self-interested, information-bound agents. We conjecture that a market economy might be an effective mechanism for achieving this goal. Therefore we decided to implement our telecommunications network management framework using economic concepts and techniques.

18.3   System Architecture

The overall system architecture consists of three layers (see Figure 18.1). The lower layer is the underlying telecommunications infrastructure. The middle layer is the multi-agent system that carries out the network management function. The top layer is the system's interface to the call request software. More details of the rationale to this design are given in Gibney and Jennings (1998). The remainder of this section concentrates on the agent layer: describing the main components (section 18.3.1) and how they relate to one another (section 18.3.2).

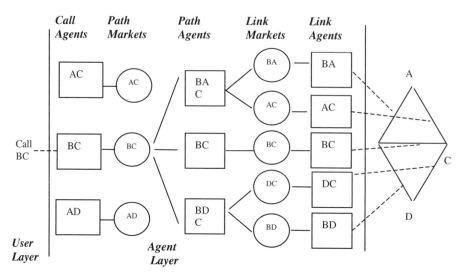

Figure 18.1   System architecture

18.3.1   The agents and their interactions

The system makes use of three agent types: (1) the *link agents* (section 18.4.1) that represent the economic interests of the underlying resources of the network; (2) the *path agents* (section 18.4.2) that represent the economic interests of paths across the network; and (3) the *call agents* (section 18.4.3) that represent the interests of callers using the network to communicate. A link agent is used for every link in the network and is deployed at the entry node for that link. A path agent is used for each logical path in use across the network and is deployed at the source node for that path. Here we use three path agents for each source destination pair. Three is a reasonable number of alternative paths across which to share a single traffic requirement (alternative static routing systems commonly using three or fewer paths). A call agent is used for each source destination pair in the network and is deployed at the source of the traffic requirement that it represents.

The agents communicate by means of a simple set of signals that encapsulate offers, bids, commitments, and payments for resources. We couple the resources and payments with the offers and bids respectively. This reduces the number of steps involved in a transaction (committing agents to their bids and offers ahead of the market outcome), and so increases the speed of the system's decision making (an important consideration in this domain). To enforce these rules the interactions between the different agent types are mediated by means of market institutions (described in section 18.3.2).

An important notion in agent technology is that agents should be proactive (i.e. able to anticipate the requirements of the environment and behave accordingly). In our system, we apply this concept to implementing a call routing mechanism that does not need to examine the network state before routing each call. Our path agents proactively determine how many calls they will be able to handle in advance, and seek to obtain the necessary resources to handle them. To be able to offer resources to callers proactively, the path agents lease bandwidth from the link agents over a period of time, paying instalments on the lease at prices agreed on the link markets.

18.3.2   Market institutions

Our system makes use of two types of market institution: at the *link market* (section 18.3.2.1), slices of bandwidth on single links (the fundamental resources of the system) are sold to path agents; at the *path market* (section 18.3.2.2), the slices of bandwidth on entire paths across the network are sold to call agents to connect calls.

18.3.2.1   Link markets

Link markets are sealed-bid double auctions. In this protocol, the auctioneer receives two sets of prices in each trading period – bids for resources from buyers and offer prices from sellers – and computes the successful trades according to a set of rules defined for the auction. A sealed-bid protocol was chosen because it provides a means to complete institutionally mediated bargaining in one shot. Therefore bargaining that would take an indeterminate time using iterated market institutions such as continuous double auctions can be completed almost instantaneously.

The resources exchanged at the link markets are the right to use slices of bandwidth on individual links which, when taken together, provide the necessary bandwidth to connect calls across paths. The link markets use a sealed-bid double auction in which buyers and sellers periodically submit bids for individual units of the resource. In our case, buyers and sellers are constrained to their roles in the market by their position in the network. Thus, path agents need to buy resources from link agents to offer services to callers. The bids and offers are ordered from high to low, and low to high respectively. There will be a range of prices for which the market will clear at which the maximum amount of resources can be traded. Buyers bidding above these prices and sellers offering below them are allowed to trade. The buyers and sellers within this group are matched randomly so that the benefits of trade are also assigned between successful agents randomly. The trading price for each given transaction is determined at random in the range between the buyer's bid and the seller's offer. Notice that this procedure implies that no buyer will pay more than their bid, and no seller will get paid less than their offer. Moreover, the procedure implies that the total surplus realised in the market, a measure of the social welfare, is maximised because the benefits of trade are distributed randomly between all successful agents.

18.3.2.2   Path markets

The path market is also a sealed-bid auction. This is because it is a critical performance requirement of the system that the allocation of call traffic to paths occur almost instantaneously (so that callers are not kept waiting for calls to be established). This means the auction protocol has to be as short as possible. As before, the most efficient protocols in this respect are the single-shot, sealed-bid types. Since we have a single caller and multiple path agents offering resources, a single sealed-bid auction is appropriate.

A buyer sending a service request message to the market initiates the auction. The auctioneer then broadcasts a request for offers to all agents able to provide the connection. All sellers simultaneously submit offers and the lowest one wins the contract to provide the connection. In this market, we experimented with two protocols: (i) the first price protocol, in which the price at which the buyer and the seller trade is that of the highest bid submitted; and (ii) the Vickrey, or second-price, auction protocol, in which the price at which the buyer and seller trade is that of the second highest bid submitted. We chose to experiment with two strategies because economic theory predicts that Vickrey auctions provide more competitive market outcomes, doing away with wasteful speculation by encouraging truth-telling behaviour on the part of the participants (Varian, 1995). However, since we are using simple adaptive agents without speculative bidding strategies, we were unsure as to whether this factor would impact the overall behaviour of the system. To test the impact of this factor on the system as a whole, we implemented the market with both protocols and empirically tested the efficiency of each (section 18.5).

18.4   Designing Economically Rational Agents

The range of potential interactions is determined by the market institutions in which the agents participate. In both of our market types, agent communication is restricted to setting a price on a single unit of a known commodity. Therefore, agents set their prices solely on the basis of their implicit perception of supply and demand of that commodity at a given time. When a resource is scarce, buyers have to increase the prices they are willing to bid, just as sellers increase the price at which they are willing to offer the resource (*mutatis mutandis* when resources are plentiful). Here, agents perceive supply and demand in the market through the success or otherwise of bidding at particular prices.

18.4.1   Link agents

A link agent has a set of n resources, slices of bandwidth capacity required to connect individual calls, that it can sell on the link market. At time t, the price to be asked for each of these units is stored in a vector $p_t = \{p_t^1, p_t^2, ..., p_t^n\}$ with the range of possible prices being zero to infinity, $p^i \in [0, \infty)$ for each member of the vector $i = 1,..., n$ and each time period t. At time $t = 0$ the prices for each unit are randomly (uniformly) distributed on $[0, H]$ where H is the initial upper limit on prices asked. When x units have been allocated, the remaining $n - x$ units are offered to the link market for sale simultaneously. Suppose that, of the $n - x$ units offered for sale in a given period t, the m units with the lowest prices are successfully sold. The prices in the vector are updated as follows:

$$p_{t+1}^i = p_t^i \qquad\qquad\qquad \text{for } i = 1 \text{ to } x \qquad\qquad (18.1)$$
$$p_{t+1}^i = p_t^i(1 + \varepsilon) \qquad\qquad \text{for } i = x+1 \text{ to } x+m \qquad (18.2)$$
$$p_{t+1}^i = p_t^i(1 - \varepsilon) \qquad\qquad \text{for } i = x+m+1 \text{ to } n \qquad (18.3)$$
$$\text{where } \varepsilon = U(0, \sigma)$$

Thus the link agent increases or decreases the price of any unit by a small amount ε after each auction. Here ε is obtained from a uniform random distribution between zero and σ (here 0.1). If previously allocated units are released by the path agent, they join the pool of unallocated units and the price vector is reordered to reflect this. This approach was chosen so that the prices of each resource on the link, taken together, should adapt to the demand on the network to carry traffic.

18.4.2   Path agents

A path agent acts as both a buyer of link resources and a seller of path resources. Their buying behaviour is detailed in section 18.4.2.1, and their selling behaviour in section 18.4.2.2. In general, path agents wish to buy resources cheaply from link agents and sell them at a profit to end consumers. To do this, they bid competitively to acquire resources that they then sell on to callers, at a price not less than that paid for them. The path agent tries to maximise its profits by adjusting its inventory and sales behaviour on the basis of the feedback it receives from the market. The mechanism by which the path agent decides what resource level to maintain is described in section 18.4.2.3.

18.4.2.1   Buying behaviour
A path agent actively tries to acquire resources (units of link bandwidth needed to connect a call), across the chain of links that it represents. It does this by placing bids at each of the link markets at which the resources it needs are sold. The agent retains a vector of prices that it is willing to pay for resources on each of the links that constitute its path. The agent's strategy is to try to equalise its holding of resources across each of those links; uneven resource holdings have to be paid for but cannot be sold on or bring in any revenue because they do not constitute complete paths. The path agent tries to maintain its resources at a level w that is discovered through hill-climbing adaptation (Russell and Norvig, 1995) to the behaviour of the market (section 18.4.2.3). The most profitable value of w is obtained by adjusting it according to changing profit during ongoing buying and selling episodes. When x units have been acquired, the path agent bids for the remaining $w - x$ units on the link market simultaneously. Suppose that, for a given link, of the $w - x$ units that the path agent bids for at time t, the m units with the highest prices are successfully acquired. The prices in the vector are updated as follows (using ε as defined previously in section 18.4.1).

$$
\begin{array}{llr}
p^i_{t+1} = p^i_t & \text{for } i = 1 \text{ to } x & (18.4) \\
p^i_{t+1} = p^i_t(1 - \varepsilon) & \text{for } i = x+1 \text{ to } x+m & (18.5) \\
p^i_{t+1} = p^i_t(1 + \varepsilon) & \text{for } i = x+m+1 \text{ to } w & (18.6)
\end{array}
$$

This price-setting mechanism was chosen because it allows the path agents to adaptively determine prices for individual link resources. The price bid for each resource should be as low as possible without failing to win the resource in the auction. Therefore the agent makes a bid for each resource that it needs separately. If a bid fails, the agent increases the price it will bid at the next auction (in order to increase its likelihood of winning the resource). If a bid succeeds, the agent reduces the price it bids for that resource in subsequent auctions (in order to avoid paying over the market price).

18.4.2.2   Selling behaviour
A path agent will offer to sell a path resource whenever an auction is announced by the path market and it has an appropriate path resource to sell. The price asked is determined

by the cost of acquiring the underlying link resources and the outcomes of previous attempts to sell. Let p_t be the price of a path resource at time t (the time of the auction) ranging from the cost to acquire the resource to infinity, $p_t = [cost, \infty)$. The first time the agent offers a resource for sale, it offers it at a price given by $p_t = cost\ (1 + \varepsilon)$ in order to sell at a profit (using ε as defined previously in section 18.4.1). Subsequently the offer price is given by:

$$p_{t+1} = \max(p_t(1 + \varepsilon),\ cost) \qquad \text{if last offer was successful} \qquad (18.7)$$
$$p_{t+1} = \max(p_t(1 - \varepsilon),\ cost) \qquad \text{if last offer was not successful} \qquad (18.8)$$

This price function ensures that the path agent never sells a resource for less than it paid to acquire it in the first place. Given its inventory level (see section 18.4.2.3), the agent attempts to maximise its income. The price bid for each resource should be as low as possible without failing to win the resource in the auction. Therefore, the agent increases the price it asks whenever it is successful, and decreases it whenever it is unsuccessful. This means the agent adapts its price to the level of competition, as it perceives it from the outcomes of previous auctions.

18.4.2.3   Inventory level

The resource levels of the various path agents determine the maximum flows available for traffic on individual paths. In our case, the system design philosophy is to have the individual paths determine their own optimal resource levels. When this is achieved, balancing the load in the network as a whole becomes an emergent property of the social interactions of the agents. Path agents act in response to the economic pressures exerted on them by their consumers, competitors and suppliers. Therefore, we choose to have path agents discover their own optimal flows by adaptation to economic conditions as they perceive them (through interactions in the markets in which they compete). Path agents are both *buyers* and *sellers* that attempt to maximise their profit through trade, where profit is the difference between their revenue (from selling path resources to callers) and their expenditure (cost of acquiring and holding on to resources). In order to maximise profit, agents must have an inventory level that is optimal for them, in the competitive environment in which they find themselves. Therefore our path agents adapt their inventory levels to the profits they earn through their interactions with the market. This is implemented by having the inventory level of individual path agents climb the hill of their profits.

In more detail, let R_t be the profit of an agent at time t and I_t be the desired resource inventory of that agent. If profit has increased since the last market interaction ($R_t > R_{t-1}$) and the last change corresponded to an increase in desired inventory level ($\Delta I_t > 0$), the new desired inventory level is increased by one resource unit; if the last change in desired inventory was negative ($\Delta I_t < 0$), then desired inventory is reduced by one unit. However, if profits have fallen, ($R_t < R_{t-1}$) and the last change was positive ($\Delta I_t > 0$), we decrease the desired inventory level; if the last change was negative ($\Delta I_t < 0$), then the desired inventory level will be increased. When decreasing the desired inventory level, the agent chooses to give up the most expensive of its link resources (that are not allocated to a call). This strategy reduces the agents' inventory rental cost by the largest amount possible in a single time step.

18.4.3   Call agents

Call agents initiate the auctions at which path agents compete by signalling that they wish to buy resources to a given source destination pair. In doing so, they set a maximum price that they will not exceed to make a call. This puts downward pressure on the offers made by path agents to provide resources across whole paths, thereby anchoring the system.

18.5   Experimental Evaluation

Our experiments were designed to test three hypotheses. First, we wanted to know whether market-based systems can compete with static routing algorithms in terms of call routing performance (section 18.5.1). Secondly, we wanted to know if our system uses the network efficiently (i.e. does it use the best routes?) whenever possible, allowing for congestion (section 18.5.2). Thirdly, we wanted to test if the system discriminates in its choice of routes between paths that would be indistinguishable from one another to a conventional static routing algorithm, without expected traffic predictions (i.e. they differ only in their proneness to congestion) (section 18.5.3).

18.5.1   Performance evaluation

In terms of performance, we sought to address two fundamental questions. First, can our market-based control system perform as well as or better than a conventional system? Secondly, what is the effect of using a Vickrey auction protocol rather than a first-price auction protocol at the path market?

In a series of experiments we tested the efficiency of our market-based control mechanism (using both first-price and Vickrey auctions as path markets) against a static routing mechanism. The static routing mechanism used a number of paths (three) between each source destination pair and routed traffic to these in order of path length, using longer paths when the shorter ones became congested. Here efficiency was measured as the proportion of calls successfully routed through the network as a percentage of the total number of calls offered. The experiment was configured to simulate a small irregularly meshed network of 8 nodes with link capacities sufficient for 200 channels. Calls arrived on average approximately every 5 seconds, were routed between a randomly chosen source destination pair, and lasted an average of 200 seconds. Call arrival and call duration were determined by a negative exponential time distribution function, with $U(0, 1)$ being a uniform random distribution between 0 and 1. The inter-arrival time between calls and the call duration were calculated using the formula $f(x) = -\beta \ln U$, where β was the mean inter-arrival time and U mean call duration, respectively. The simulation was allowed to run for 20,000 seconds in each case. The traffic model was chosen to simulate a realistic call arrival rate and duration. The network dimensions where chosen to reflect a small network under heavy load.

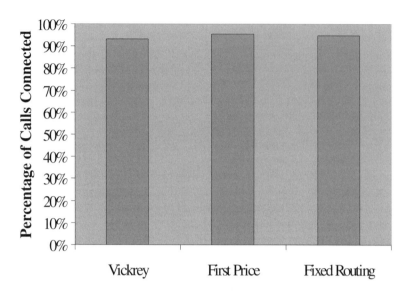

Figure 18.2   Performance of market-based and static routing

These results show that similar levels of performance are obtained using the market-based control mechanism (95.4% of calls connected) and static routing (94.8% of calls connected) (Figure 18.2). It is interesting to note that, contrary to our original hypothesis, using a Vickrey auction for the path market did not improve upon the results obtained using a first-price auction. One possible reason for this is the way in which the path agents in our system adapt their pricing strategy to market outcomes. Vickrey auctions are designed to make the markets more efficient by making counter-speculation between competing agents wasteful. However, with simple adaptive agents being used here such speculation does not occur, so the effect should not be significant.

The ability of our system to perform as well as a static routing mechanism should be taken as a positive result. As well as matching the performance of conventional routing techniques, our market-based approach has a number of distinct advantages for network operators and users. First, it provides an architecture that is open to deployment in multi-enterprise environments without the inefficiency of static internetworking policies at sub-network interfaces. Secondly, our system is scalable in that no agent has to know the address of a significant number of peer agents or possess a map of the entire network. Thirdly, our system allows a much quicker response to call requests because the call routing process does not need to obtain information from the wider network at call set-up time. In our case a call request can be processed and the call dispatched to a path (or refused) solely on the basis of information present at its source. This is achieved by having path agents that proactively determine their capacity to handle traffic, rather than waiting for call requests before processing.

18.5.2   *Resource utilisation efficiency*

In addition to the raw connection performance, it is important to know how effectively the network's resources are used. This is important because both the callers and the system benefit from routing calls through the shortest path when one is available. Shorter call routes use fewer system resources than longer ones, and they provide a service with less delay to end users. To assess our system's performance, we analysed the relative percentages of calls that were assigned to first, second and third best paths (by the number of links which make up the path). The data (Figure 18.3) shows that the majority of calls were routed through the most efficient routes: 61% using the most efficient route, 25% the second best route and 15% the third best. Thus not only does our system route most of the calls it is presented with, it also makes the most efficient routing choices.

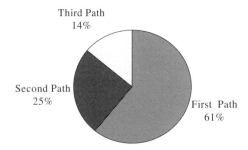

Figure 18.3   Utilisation of first, second and third shortest paths by market-based control

18.5.3   *Congestion discrimination*

One of the claims we made for market-based approaches is that good system-level choices can emerge from local choices that are influenced by information about the social context (obtained through interaction). To explore this hypothesis, we examined the performance of our system in cases that are indistinguishable from a local perspective. Thus, we focused on source destination pairs where all the routes are of equal length. In such cases, an alternative routing mechanism cannot decide between these paths without some notion of congestion through the whole network, which cannot be calculated and propagated in real time (recall the discussion of section 18.2). Alternative routing mechanisms can assign traffic to paths probabilistically, so that statistically, over time more traffic is routed to less congestion-prone paths. However, this method is dependent on the accuracy of past measured traffic as a predictor of future traffic patterns. With our approach, we believe such discrimination should emerge from the competitive nature of the market place. The reason for this hypothesis is that while path agents for paths of equal route length have to obtain the same number of resource slices, some have to obtain the more congestion-prone of those slices. By definition, the more congestion-prone resource slices are traded in the more competitive (and hence more highly priced markets). All other things being equal, the profitability of selling these paths will be lower because of the higher costs. With lower profitability comes a lower inventory level and fewer calls being routed via that path.

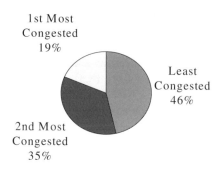

1st Most
Congested
19%

Least
Congested
46%

2nd Most
Congested
35%

Figure 18.4   Utilisation of paths in order of congestion by market-based control

We wanted to test whether the market-based control mechanism is able to discriminate between paths on the basis of congestion cost in real time. In order to investigate this, we examined the source destination pairs in our network configuration (nine of them) for which all (three) known paths consisted of an equal number of links. We plotted the percentage of calls routed to each of the three paths in (reverse) order of congestion and took the average of these values across all nine source destination pairs (Figure 18.4). Our results clearly show that the market mechanism is able to distinguish between congestion costs entailed in routing across paths of otherwise equal length and assign calls to the least congested path most of the time.

18.6   Conclusions

We have described the design and implementation of a market-based system for call routing in telecommunications networks. Our system performs comparably with a static routing approach in terms of the percentage of the calls that are connected. However, from an architectural point of view, the market-based approach represents an improvement on static, centralised systems for a number of reasons. First, it provides a platform for implementing network traffic management in a multi-enterprise internetwork. Secondly, it does not rely on a centralised controller to compute network reconfigurations, making the network management function robust to failure. This means that the system described in this work is architecturally more robust than an equivalent network with a centralised control mechanism. It is important to distinguish between this sense of robustness, and robustness as an empirical measure of the performance of the system in the event of the failure of a component of the system (or the agent process managing it). The performance of a decentralised control system degrades with the loss of controlling processes, while a centrally controlled one would be without control if the controlling process were to be lost. The study of the degradation of control efficiency in the system described here has been left to future work. Thirdly, no agent needs to know of the existence of more agents than there are links in the paths of the network (making the agent acquaintance databases compact and the whole system more scalable). Fourthly, there is no requirement to test the network state at call set-up time, making the call set-up procedure faster and more robust. Fifthly, the cost of each call to the network and the proportion of that revenue owing to each of the enterprises involved in carrying that call can easily be computed from information available to the user terminal equipment when the call is made, thus making call charging more efficient.

The results presented in this paper show that our market-based system performs the call routing and network management tasks adequately. However, the function used to determine the inventory level of path agents is quite simple and responds reactively to burstiness in the call arrival rate; this may be inducing unwanted oscillation in the path inventory parameter which may be adversely affecting performance. We intend to experiment with this function, and the parameters that govern its behaviour, to determine the impact of our choices on the performance of the overall system, and to see if that performance can be improved upon.

Acknowledgements

This work was carried out under EPSRC grant No. GR/ L04801.

18.7   References

Appleby, S. and Steward, S. (1994) Mobile Software Agents for Control in Telecommunications Networks. *BT Journal of Technology*, **12(2)**, 104–113.

Bertsekas, D. and Gallager, M. (1987) *Data Networks*. Prentice Hall International.

Cisco Systems Inc. (1999) *A Primer for Implementing a Cisco Virtual Private Network (VPN)*, Cisco Systems Inc.

Clearwater, S.H. (Ed.) (1996) *Market-Based Control: A Paradigm for Distributed Resource Allocation*. World Scientific Press.

Gibney, M.A. and Jennings, N.R. (1998) Dynamic Resource Allocation by Market-Based Routing in Telecommunications Networks, in *Proceedings of Intelligent Agents for Telecommunications Applications*, pp. 102–117, Springer-Verlag.

Goldszmidt, G. and Yemini, Y. (1998) Delegated Agents for Network Management. *IEEE Communications Magazine*, **36(3)**, 66–70.

Hayzelden, A.L.G. and Bigham, J. (1998) A Heterogeneous Multi-Agent Architecture for ATM Virtual Path Network Resource Configuration, in *Proceedings of Intelligent Agents for Telecommunications Applications*, pp. 45–59, Springer-Verlag.

Jennings, N.R. and Wooldridge, M.R. (1998) *Agent Technology Foundations, Applications and Markets*. Springer-Verlag.

Kershenbaum, A. (1993) *Telecommunications Network Design Algorithms*. McGraw-Hill International Editions.

Russell, S.J. and Norvig, P. (1995) *Artificial Intelligence: A Modern Approach*. Prentice Hall, Inc.

Schoonderwoerd, R. Holland, O.E. and Bruten, J. (1997) Ant-like Agents for Load Balancing in Telecommunications Networks, in *Proceedings of the 1ˢᵗ International Conference on Autonomous Agents*, ACM Press.

Schwartz, M. and Stern, T.E. (1989) Routing Protocols, in C.A. Sunshine (Ed.), *Computer Network Architectures and Protocols (2rd Edition)*, Plenum Press, New York/London.

Stallings, W. (1997) *Data and Computer Communications*. Prentice Hall International , Inc.

Varian, H.R. (1992) *Microeconomic Analysis (3rd Edition)*, W.W. Norton & Company Inc.

Varian, H.R. (1995) Mechanism Design for Computerised Agents, in *Proceedings of the 1995 Usenix Workshop on Electronic Commerce*.

Wellman, M. (1996) Market Oriented Programming: Some Early Lessons, in S.H. Clearwater (Ed.), *Market-Based Control a Paradigm for Distributed Resource Allocation*, pp. 74–95, World Scientific Press.

19

IN Load Control Algorithms for Market-Based Multi-Agent Systems

A. Patel, J. Barria, J. Pitt

19.1   Introduction

The number and variety of services offered by telecommunication systems has increased dramatically in recent years; this has been driven in part by technological advances and the use of *Intelligent Networks* (INs). This trend has encouraged the greater use of telecommunication services and network bandwidth and has therefore led to greater network profitability. During peak periods of resource utilisation, an IN can become overloaded with service requests, which can lead to a degradation in the *Quality of Service* (QoS) provided by the network. For example, user calls may be dropped due to buffer overruns, thus significantly reducing throughput, and consequently increasing the total call set-up time. For an IN to be able to satisfy these peak demand levels, while maintaining a high level of QoS, a load control algorithm must be applied.

An IN is an overlay network to telecommunication systems that facilitates greater service sophistication and complexity. An example of an IN configuration can be seen in Figure 19.1. Users of the IN can access IN services through the *Service Switching Point* (SSP). When an SSP receives an incoming request, it has to establish which *Service Control Point* (SCP) is capable of providing the service requested and then determine if it has the right to access that SCP. Once a link has been established, the SCP runs the *Service Logic* (SL) program code that represents the service requested. The SL code may make use of additional IN resources, such as *Service Data Points* (SDP), for database access, and *Intelligent Peripherals* (IP), for stream data access. A *Service Management System* (SMS) is introduced to manage the IN services.

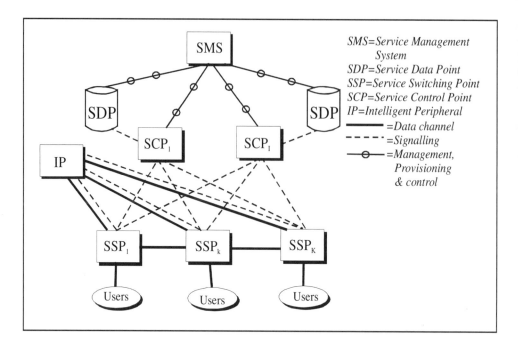

Figure 19.1   An example IN configuration

Currently IN load control algorithms are defined by their throttling mechanism, how to obtain throttling parameters and to which IN architecture they apply. A throttling mechanism defines how to restrict the incoming calls in order to alleviate overload conditions. Potential forms of throttling mechanism include *window-based* and *rate-based* mechanisms. In a window-based mechanism, such as the mechanisms defined in Pham and Betts (1994) and Bedoy et al. (1998), a limit is set on the maximum number of concurrent connections allowed; in a rate-based mechanism, for example the algorithm used in Smith (1995), a limit is set on the maximum number of accepted connections per second. In these mechanisms, the parameters for the throttling mechanism are found by monitoring the performance of the system and incrementally adjusting the parameters to find the appropriate throttling level. This can potentially lead to slow response or even oscillations in the algorithm. Alternatively, the parameters for throttling may be obtained using an optimisation algorithm (Kihl and Nyberg, 1997; Lodge, Botvich and Curran, 1998).

Many of these algorithms are based on a single-service IN architecture, that is, when a service can only be performed by a single SCP. Researchers have investigated *Advanced Intelligent Networks* (AIN) where each service is made available from multiple SCPs (Kawamura and Sano, 1996; MARINER, 1998). However, these algorithms require a centralised controller which can act as a single point of failure for the entire network.

Recently, market-based multi-agent systems have been successfully applied to distributed resource allocation problems (Clearwater, 1996; Yamaki, Wellman and Ishida, 1996). In this chapter, we develop a resource allocation framework for the distributed IN load control problem. We propose a price-oriented algorithm and a resource-oriented algorithm to find the general equilibrium of this resource allocation problem.

In section 19.2, we provide a background on market-based algorithms. In section 19.3, we develop the resource allocation problem that represents the distributed IN load control problem. In section 19.4, we present a price-oriented and a resource-oriented solution to the distributed IN load control problem. We present simulation results for the two models in section 19.5. Finally, we conclude in section 19.6.

19.2   Algorithms for Control of SCP Overloads

An IN can be viewed as a set of distributed network resources. Any one of these resources can become *overloaded* when the number of simultaneous requests to use it exceeds the design capacity. Unless a congestion control is in place, this results in: (i) a significant increase in the call-processing delay; and (ii) a significant drop in the network throughput.

Future INs will see an increase in the number and variety of services that are being provided. This means that they will, in particular, suffer from focused overloads (Atai and Northcote, 1996). In these cases, extraordinary events such as media-stimulated mass call-in events, e.g. tele-voting or natural disasters, can cause an unexpected sharp increase in network usage which leads to the overload of network resources.

Within an IN, the individuals who own, operate and use the network represent different network roles. Each of these roles has a different objective and overload affects them in different ways:

1. The *users* of IN services are assumed by many researchers to want fair access to IN services with fair access being defined in many different ways.
2. The *equipment manufacturers* of the network resources want their equipment to operate with the highest possible efficiency and process the maximum number of calls possible.
3. The *network operators* want to maximise the amount of money generated by the network, i.e. the call revenue, without damaging its public image, which represents long-term and sustainable revenue.
4. The *service providers*, who vie for customers, would want the QoS supplied by the network to be fairly distributed between the service providers.

An IN load control algorithm consists of: (i) a *throttling mechanism*, which defines how and which of the incoming calls will be rejected, in order to reduce the overload within the network; (ii) a *triggering mechanism*, which defines when call throttling commences; (iii) *determining the parameters*, which defines how the parameters for the throttling mechanism are obtained; and (iv) *alleviating the throttling*, which defines when the throttling mechanism is released.

19.2.1   Throttling mechanisms

Potential throttling mechanisms include *call gapping* (CG), *window*, and *percent thinning*. Throttling mechanisms are divided into two groups: these are the rate-based mechanisms and window-based mechanisms. An example of a rate-based mechanism is the CG scheme.

In this scheme, each SSP holds two variables: the gap, g, and the duration, d. The SSP is allowed to send one connection request in the interval g for the next d seconds. Another rate-based mechanism is the percent thinning control; in this scheme the SSP uses a parameter ρ which determines the probability that a call is accepted. Finally, the *input-rate selection* scheme defines the maximum number of service requests per unit of time to any particular SCP from each SSP. This mechanism does not explicitly specify which calls to reject but it is left to the discretion of the SSP exactly which calls to throttle.

Window-based throttling mechanisms limit the total number of ongoing service requests. In a window-based scheme, each SSP uses two variables the first is the current window size, W, and the second is the current number of active connection requests, C. All connection requests are rejected when $C=W$. This sort of scheme is more difficult to implement since additional signalling would be required if an ongoing call has to be dropped.

The choice of throttling mechanism can determine the effectiveness of the overall algorithm. The CG mechanism automatically rejects calls that are too close together. This means that larger sources are throttled more than smaller sources for the same parameter value. The percent thinning mechanism randomly determines whether to throttle a connection or not - this method is fairer to larger sources; however, if there is a sudden increase in arrival rates, this mechanism may still allow the SCPs to become overloaded.

The window mechanism does not enforce a time between accepted calls. Services are handled on a first-come first-serve basis; if there is a sudden increase in the incoming call rates then the window will fill with calls from the first few calls and later calls will be throttled.

19.2.2 Triggering mechanism

A throttling mechanism may need to be triggered when an overload occurs or may be always active. A mechanism which is always active needs more processing time during underload periods; however, this form of mechanism may be able to react faster to a sudden overload situation.

A mechanism that is triggered by an overload is known as reactive. In these mechanisms, the SCP performance is monitored for signs of overload, for example, by using average processor utilisation (Smith and Northcote, 1998). When one or more of these performance measures surpasses a threshold value, throttling is activated.

A related issue is how to alleviate throttling. An inappropriate choice of triggering points to alleviate throttling can lead to oscillations in the control (Pham and Betts, 1994). A simple method of alleviating throttling is to stop throttling after a given duration, e.g., the parameters in the CG mechanism. This is ineffective and can lead to oscillations in the control (Smith, 1995). SCPs find it difficult to determine when the overload has passed since they only see the incoming call rate after it has been throttled at the SSP. The percent-thinning throttling mechanism overcomes this. If ρ is the thinning parameter and the arrival rate of calls at the SCP is λ, then the arrival rate at that SSP is λ/ρ. This allows the SCP to make more informed choices about the throttling parameters.

19.2.3   Determining the throttling parameters

The parameters may be calculated at the SSP or the SCP. If they are obtained only by the SSP then this is known as a non-interactive algorithm. Alternatively, in an interactive algorithm the SCP obtains the parameters and passes them to the SSP. An example of a non-interactive algorithm is described in Pham and Betts (1994). The SSPs use the window throttling mechanism and monitor the acknowledgement messages received from the SCPs; when a time-out occurs the SSP assumes the SCP is overloaded and reduces the window size.

An interactive algorithm known as Automatic Call Gapping (ACG) is described in Smith (1995). In this form of algorithm, the SCP uses monitoring information to determine appropriate parameters. The mapping between monitoring information and parameter values is not obvious and a dynamic search may be used.

19.2.4   Node-level algorithms

A node level algorithm is one that uses only locally available information to determine throttling parameters. Examples of these algorithms are ACG (Smith, 1995) and the window-based algorithm (Pham and Betts, 1994). Comparative studies of node-level algorithms can be found in Sidi, Lui, Cidon and Gopal (1993), Pham and Betts (1994) and Kihl and Nyberg (1997). They show that the most responsive algorithms are interactive rate-based ones. Node-level algorithms have several disadvantages:

- They cannot optimise the network for a given criterion, which results in unbalanced throttling across the network and fairness suffers as a result.
- As nodes act to protect themselves against overload, other nodes become overloaded by calls diverted to them, thus propagating the overload across the network.
- Most of these algorithms are defined for the case when one SCP provides a single-service logic and extending these to the multi-service case is not trivial.

These problems occur because the actual information needed to make an optimal choice of throttling parameters is distributed amongst the various nodes in the network.

19.2.5   Network-level algorithms

Recently, several algorithms have been proposed that use network level information to determine the throttling parameters, e.g. explicitly using incoming rates at the SSP together with the maximum processing capability of the SCP; for example, see the algorithms described in Kawamura and Sano (1996), Kihl and Nyberg (1997), Hac and Gao (1998) and Arvidsson, Jennings and Angelin (1999).

The algorithms proposed in Kawamura and Sano (1996) and Arvidsson et al. (1999) use a centralised controller that uses network utilisation information to determine the exact cause of an overload and the appropriate throttling parameters. Other methods have also been proposed that involve exchanging load information between nodes (Hac and Gao, 1998). Network-level solutions suffer from several drawbacks which include:

- The algorithms can be more complex than node-based ones.
- The communication overhead is much higher than with node-based schemes.
- The algorithm for optimising the network is typically run as a centralised system, which reduces the reliability and responsiveness of the algorithm.

While the network-level algorithms can optimise the network for any given level of load, the need for a centralised controller can create a bottleneck in the network and can be disastrous should the controller fail. For this reason, we investigate the use of market-based schemes to provide a distributed solution to the problem of IN load control.

19.3   Market-Based Multi-Agent Systems

Market-based multi-agent systems (MAS) have been applied to the field of distributed resource allocation (Clearwater, 1996; Yamaki et al., 1996). The use of multi-agent systems for distributed applications has some inherent advantages that make them well suited to the application of distributed IN load control; these include:

- The problem is geographically distributed.
- The network can be easily mapped into a society of autonomous components.
- Agents conceptually provide a high level of control which can make use of network-level information which is known to give better performance.

In a market-based multi-agent system, *economic agents* are used to represent individuals in a fictional computational economy. The economy consists of commodities, which represent the resources to be traded, and agents, which represent the individuals with a stakeholding in the problem to be solved.

There are various types of agents available. Each agent will try to trade with other agents to best suit its preferences and behaviour. The preferences of the agent in the economy are modelled by a utility function – if the individual represented is a *consumer* – or a production function – if the individual represented is a *producer*. In addition to this agent, auctioneers are introduced into the economy, which control the allocations of each commodity. In order for the agents and auctioneer to successfully exchange commodities they must participate in an *auction* for each commodity. The *auction protocol* defines the algorithm by which agents can adjust their allocations of commodities. The choice of auction protocol will depend to an extent on the application considered, but different protocols will possess different properties. There are many different auction protocols available (Friedman and Rust, 1993; Wellman, Walsh, Wurman and MacKie-Mason, 1998) however, we have chosen to investigate the algorithm known as WALRAS (Wellman, 1993) and the resource-oriented Newton method (Ygge and Akkermans, 1997).

Market-based approaches can be divided into two broad categories; these are price-oriented algorithms and resource-oriented algorithms:

- In a price-oriented algorithm, commodities are associated with prices. Agents exchange commodities based on the current price until the level of supply and demand is met for the commodity. An auctioneer is used to adjust the price of commodities. An example of such an algorithm is the WALRAS algorithm (Wellman, 1993).

- In a resource-oriented algorithm, individual agents know how much they are prepared to pay for an additional unit of allocation. All the agents send their bids to an auctioneer who adjusts the allocations to equalise the price paid by all agents. An example of such an algorithm is the resource-oriented Newton method (Ygge and Akkermans, 1997).

The price- and resource-oriented alternatives have different properties, some of which are identified below:

- The final allocations of commodities are found by iteratively adjusting the current allocation in a way that will be more beneficial to the agent participating in the auction. The allocations for price-oriented algorithms may not be feasible until an equilibrium is reached; since the individuals within the economy decide their own demand level, they may choose a demand level that surpasses the allocation available for that commodity. For the resource-oriented case, it is always possible to find a feasible solution, since the auctioneer decides and adjusts the allocations of commodities and can therefore always adjust them to be feasible at every iteration.
- In the resource-oriented case, since the auctioneer adjusts the allocations, it reduces the scope for distributed decision making. In the price-oriented case, the auctioneer can be made very simple while the demand bids are generated by the agents. This can help to distribute the computational cost of finding a solution and also allows scope for a wide variety of different bidding strategies to be adopted by the individuals.
- Production is the process of converting one commodity into another. The production of commodities is important for the application of distributed IN load control. In a price-oriented algorithm, general production is possible. However, in a resource-oriented case, production is much more difficult to achieve since auctioneers need to ensure feasibility at every iteration. Ensuring feasibility at every iteration and finding the optimal solution is much more difficult than for an economy where production of goods is prohibited.

Alternative auction protocols exist which are hybrids of the price- and resource-oriented approaches, such as double auction markets (Friedman and Rust, 1993). These have been used to try to eliminate the various problems associated with the price- and resource-oriented alternatives. However, we have chosen to investigate algorithms based on the price- and resource-oriented alternatives, which represent opposite ends of the spectrum of possible solutions, with the aim of investigating the specific issues within market-based systems that make them either well suited or not well suited to the application of distributed IN load control.

19.3.1 Price-oriented market-based approaches

For price-oriented algorithms, agents send demand bids to auctioneers which in turn return the price level for that commodity. These demand bids can be generated in various ways according to the type of agent being identified. There are two types of agents used in this paper these are the *consumer agent* and the *producer agent*. For the case of a consumer agent, consumer k generates its demand bid by maximising its utility function, $u_k(x)$, given an income constraint. This behaviour can be defined using the following maximisation problem:

$$Max_x \; u_k(x) \quad s.t. \; \mathbf{p}.\mathbf{x} \leq \mathbf{p}.\mathbf{e}, \tag{19.1}$$

where $\mathbf{x}$ is the commodity allocation vector and $\mathbf{e}$ is the initial endowment of commodities.

For producer agents, a demand bid is generated by maximising the profit made from the production. In this chapter, we consider the simple case where each producer has a single input commodity, of which its allocation is x_k, and a single output commodity, of which its allocations is y_k. The producer behaviour is then defined by:

$$Max_x \; p_y y_k - p_x x_k, \tag{19.2}$$

where p_y is the price of the output commodity and p_x is the price of the input commodity.

The price-oriented algorithm being investigated in this paper is the WALRAS auction protocol (Wellman, 1993); all agents wishing to participate in a particular auction need to send an entire demand curve, i.e. a curve of demand bids as a function of the price of that commodity assuming that all other commodity prices remain unchanged. Given the demand curve of the kth agent as $D_k(p)$, the auctioneer calculates the price of the commodity using the following definition:

$$p_j = p_j^* \quad s.t. \; \sum_k D_k\left(p_j^*\right) = 0. \tag{19.3}$$

The auctioneers send the updated prices back to the agents which can then determine a new demand level given the various new prices from the auctioneers. The original demand curves were based on the premise that the price of all commodities remained the same. If this is not the case, then the new demand level set by the agents may exceed the level expected by the auctioneer and hence this is an infeasible allocation of commodities. The auction protocol is repeated until a feasible allocation is found.

19.3.2 Resource-oriented market-based approaches

In resource-oriented algorithms, the agents send price bids to the auctioneers. This price bid is the price that an agent is prepared to pay for a marginal increase in its allocation of a particular commodity. The price bid for commodity j by agent k, $p_{j,k}$, can be generated from the utility function using the following relationship:

$$p_j = \frac{\partial U_k(\mathbf{x})/\partial x_j}{\partial U_k(\mathbf{x})/\partial x_J}, \tag{19.4}$$

where J is the total number of commodities in the economy.

The objective of the auctioneer is to adjust the allocations of commodities in such a way that the price bids from all the agents for a particular commodity are equalised. This is equivalent to the auctioneer behaviour defined by equation (3). The resource-oriented algorithm being investigated in this chapter is the resource-oriented Newton method. This algorithm uses a single auctioneer to adjust the allocation of all the commodities at once. Further details of the algorithms may be found in Ygge and Akkermans (1997).

Although the resource-oriented case produces a feasible result at every iteration, this allocation may be far from optimal. For this reason, we assume that regular auctions are held for the commodities. During these auctions, the utility function and the preferences of the individual agents are held constant. Once the results of the auctions are known, the allocation of commodities at the end of these auctions reflects the overall allocation of real-life resources.

In an IN environment, this approach may lead to unsatisfactory results since this environment is dynamically changing. If the time taken for the auction to be completed is too long, so that the preferences of the individuals have changed significantly during this period, then the results of the auction may be far from optimal. For this reason, we introduce a real-time deadline to find a solution to the auction. The effect of this on the two different algorithms presented herein is discussed in the sections describing the algorithms.

19.4   IN Load Control Using a Market-Based Solution

In the studied network configuration, which is similar to the network shown in Figure 19.1, we assume there are a total of K SSPs, I SCPs, and J SL types. We assume that the SDPs and IPs are infinite server queues. We study an IN model in which any service can be provided by any SCP. We first develop the resource allocation framework that represents the IN load control problem. To achieve this we use an *input rate selection* throttling mechanism. This is a rate-based throttling mechanism where the parameter x_{ijk} is defined, which specifies the maximum number of call requests of SL type j that SSP k is allowed to send to SCP i per second. The objective of the market system is to determine the values of the parameters x_{ijk}, that is, to allocate SL access rights to the SSPs.

Each SL type provided by the network is defined by a single parameter, s_j, which represents an exchange rate between the cost of SL type j and raw processing capacity at an SCP. The average processor utilisation is then defined by:

$$\rho_i = \sum_{j=1}^{J}\sum_{k=1}^{K} s_j x_{ijk}. \tag{19.5}$$

Note that this function effectively implies that a production of goods is required to convert the distributed IN load control problem into a market-based problem. This is because the raw SCP processing capacity needs to be converted to services, with the exchange rate between the services determined by s_j. In designing the algorithms, consideration needs to be taken as to how this production is to be implemented. Particularly in the resource-oriented case, which is not as well suited to economies in which production exists, compared to the price-oriented schemes.

From an SCP's point of view, its raw processing capacity needs to be allocated to the individual SSPs. The processing capacity of SCP i is given by C_i. A constraint placed on the system is then to limit the utilisation of each SCP below its capacity:

$$\rho_i \leq C_i. \tag{19.6}$$

Our objectives in defining these alternative algorithms are not to define two equivalent models, but rather to use the properties of the individual approaches to create two models for the distributed IN load control problem.

19.4.1   A price-oriented alternative

The computational economy for the price-oriented algorithm is shown in Figure 19.2. This model is based on the model proposed in Yamaki et al. (1996), with two differences: (i) the model is not restricted to a single bandwidth and future bandwidth, but rather is mapped to fit the view of the physical IN network with multiple-bandwidth and future-bandwidth markets; and (ii) the consumer agents are not tied to the structure of the producer agents, which means that each SSP is free to choose in which markets to bid, and each SCP is free to choose which services to provide (again this is more representative of a true IN structure).

In Yamaki et al. (1996), a future-bandwidth commodity is introduced to provide the agents with an incentive to trade allocation of commodities. We have introduced multiple future-bandwidth commodities labelled FBW_i to provide the same trading capability. This also means that there is no centralising factor in this model, which makes it more resilient to agent failures. The circles represent the different agents while the rectangles in the middle represent markets where goods can be exchanged. The network goods can be divided into two groups: (i) the *bandwidth* group, labelled BW_i and FBW_i, represent markets for bandwidth in the current and next time slot; and (ii) the *SL* group, labelled $SL_{i,j}$ and FSL_i, represent markets for current and future SL type j from SCP i. The directed edges indicate the flow of commodities in the economy.

19.4.1.1 Network demand
The SSPs are modelled as utility-maximising consumers, the behaviour of which has already been described. In this research, we investigate the use of the CES (Constant Elasticity of Substitution) preference model. This FBW_i model has a demand function that is easy to parameterise and hence minimises communication between agents and the auctioneers; however, in principle, any general utility function may be used. We denote the variables, x_{ijk} and xf_{ik} as the allocations of SL type j from SCP i to SSP k for the current and next time slot. Similarly, xb_{ik} and xfb_{ik} represent the allocation of bandwidth and future bandwidth of SCP i held by SSP k. The CES utility function is then given by:

$$u_k(xb_{ik}, xfb_{ik}, x_{ijk}) = \left\{ \sum_{i=1}^{I} [\alpha_{ki} xb_{ik}]^{\frac{\sigma}{\sigma-1}} + \sum_{i=1}^{I} [\alpha f_{ki} xfb_{ik}]^{\frac{\sigma}{\sigma-1}} + \right.$$

$$\left. \sum_{i=1}^{I} [\beta f_i xf_{ik}]^{\frac{\sigma}{\sigma-1}} + \sum_{i=1}^{I} \sum_{j=1}^{J} [\beta_{ij} x_{ijk}]^{\frac{\sigma}{\sigma-1}} \right\}^{\frac{\sigma-1}{\sigma}}. \tag{19.7}$$

The α, αf, β and βf coefficients dictate the relative preferences for commodities, while the σ parameter represents a global substitution parameter. In this form of model it is important to estimate appropriate values for these parameters, in order for the system to converge successfully and to a desirable final solution. This is because the desired parameter values depend on the total bandwidth and future bandwidth available in the network. In this case, appropriate values may be determined through offline simulations, i.e. the parameters may need to be *tuned*.

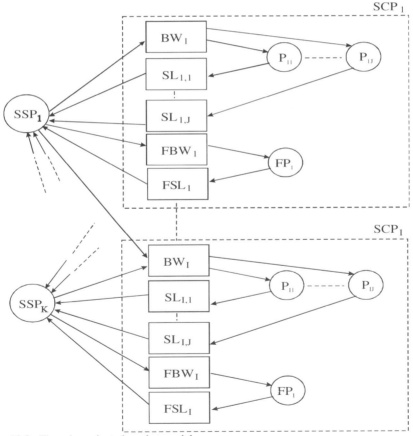

Figure 19.2 The price-oriented market model

19.4.1.2   Network supply

The SCPs are modelled as a set of producers labelled P_{ij} and FP_i. These convert the raw processing bandwidth, from the BW_i and FBW_i markets, into allocations of SL sold at $SL_{i,j}$ and FSL_i markets. The objective of the producers is to relate the cost of converting raw SCP processing bandwidth into actual SL. We define the allocation of bandwidth as x_{ij}; this is the input to the production process while the output is defined as y_{ij}. The relationship between input and output is:

$$y_{ij} = Q_{ij}\left(1 - \frac{1}{1 + x_{ij}}\right). \tag{19.8}$$

The behaviour of these agents has already been defined above and is to maximise the profit made from the production of the output commodities.

19.4.1.3   Finding an auction solution

The auction of commodities needs to finish a fixed time after the auction start time. Since this is a price-oriented algorithm, it may be the case that the solution cannot be found before this time, and hence the allocations at that time may be infeasible. In this case a heuristic algorithm, similar to the proportion algorithm proposed in Ygge and Akkermans (1997), is used to allocate a feasible – if not an optimal – allocation of resources.

19.4.2 A resource-oriented alternative

The characteristics of the distributed IN load control problem imply that production is needed to convert the raw SCP processing bandwidth into allocations of SL. However, a resource-oriented approach is not well suited to this form of economy and finding a feasible allocation at every iteration may become a complex task. To solve this problem we introduce a supplier – a consumer that is endowed with all of the commodities. Figure 19.3 shows the resource-oriented model.

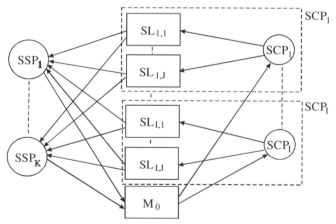

Figure 19.3   The resource-oriented market model

There are two types of markets in this economy: (i) the market labelled $SL_{i,j}$ represents the market for trading allocations of SL type j from SCP I; and (ii) the market labelled M_0 represents a monetary commodity that the agents use to trade with suppliers for allocations of SL.

In this model, the M_0 commodity is similar to the future-bandwidth markets in the price-oriented model – it is required to give the agents an incentive to trade with each other. In fact, it has been shown that using this commodity maximises the sum of the utility functions of all the agents (Bikhchandani and Maner, 1997).

19.4.2.1 Network demand

As with the price-oriented case, we model the SSP agents as consumers, with a utility function of an exponential form:

$$u_k(x_{1k},...,x_{1k},m_{0k}) = \sum_{j=1}^{J} R_j \lambda_{j,k} \left\{ 1 - e^{-x_{jk}/\lambda_{j,k}} \right\} + m_{0k}, \tag{19.9}$$

where x_{ik} is the total allocation of SL type i to SSP k; $\lambda_{j,k}$ is the arrival rate of calls of service type j to SSP k; R_j is the network revenue generated by a successful call of type j; and m_{0k} is the allocation of commodity M_0 to SSP k.

Unlike the price-oriented case, this model of utility has almost no parameters that are undefined, which means a tuning phase is not required for this model.

19.4.2.2 Network supply

The utility function of the supplier is a function of Z_i, which is the unallocated bandwidth at SCP i. The utility function used is:

$$u_i(Z_i) = a_i \log Z_i + m_{0i}, \tag{19.10}$$

where a_i is a gain coefficient and m_{0i} is the allocation of M_0 held by SCP i.

This algorithm can be implemented using the Newton method for resource-oriented systems presented in Ygge and Akkermans (1997).

19.5   Evaluation of the Algorithms

In our evaluation of these algorithms we use a network configuration consisting of 8 SSPs, 4 SCPs and 3 SL types. The services used are a Virtual Private Networking service, a Ringback service and a Call Forwarding service. In the first case we investigate the ability of the two algorithms to effectively control the traffic load. We adjust the capacity of the SCPs such that the normal incoming load is 35% of system capacity. We then apply an overload to the system, with an incoming load of 150% of system capacity for a period of 600 seconds. Our target is to maintain this load at 90% of system capacity. The results can be seen in Figure 19.4.

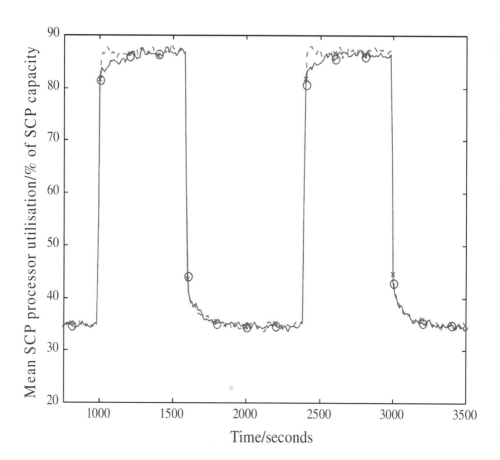

Figure 19.4   Overload performance of the price and resource-oriented (dotted lines) algorithms

In the price-oriented case, it is important the system reaches a general equilibrium solution, since this is the only point where the allocations of SL for all the SCPs are feasible. The number of iterations taken for the algorithm to converge to a solution is very important in determining the expected time between auctions and the effectiveness of the algorithm in tracking changes in traffic patterns. Figure 19.5 shows the number of iterations required to converge as a function of the incoming load.

In evaluating the performance of the resource-oriented algorithm, we investigate the effect of the parameter from equation (10) on the solution. This parameter provides a trade-off between the convergence speed and the efficiency of the system. Figure 19.6 shows the number of iterations required to converge and the inefficiency of the allocation, measured as the percentage of available bandwidth not allocated to any SSP. This figure is measured by applying a fixed incoming load of 150% of system capacity and adjusting the parameter for all the SCPs simultaneously.

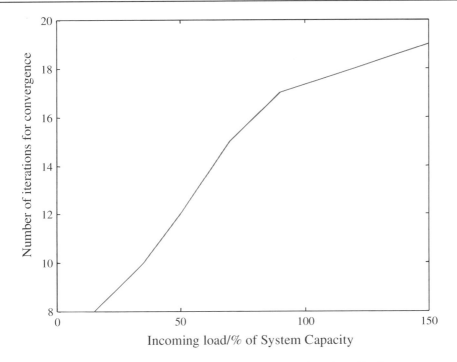

Figure 19.5 Convergence time of price-oriented algorithm as a function of load

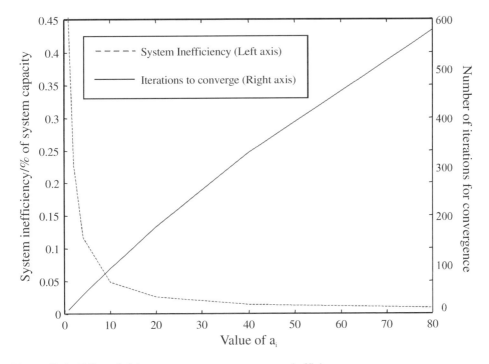

Figure 19.6 Effect of alpha parameter on convergence and efficiency

19.6   Conclusions

Many different market-based schemes for distributed resource allocation have been proposed in the literature (e.g. Clearwater, 1996; Yamaki et al., 1996). In this chapter, we studied a market-based resource allocation framework for implementing the distributed IN load control problem. We proposed two alternative approaches, one price-oriented and the other resource-oriented, each with its own characteristics. The objective of these algorithms is to allocate SL from the SCPs to the SSPs in order to implement the input rate selection throttling mechanism. The algorithms were developed with the view of implementing them in a multi-agent system.

Preliminary results show that both of the models can successfully control the incoming load to within the required levels. However, the two algorithms possess different properties. For the resource-oriented case, which is less suited to economies in which production exists, the number of iterations required to find the solution is much higher thanfor the price-oriented algorithm. However, since in this model a feasible set of allocations is available at every iteration, the algorithm can easily allocate resources once the auction time has expired. We have also investigated how the gain parameters of the suppliers affect the overall system and we have identified a trade-off between the efficiency of allocation and the speed of convergence to the optimal solution.

The price-oriented algorithm can allocate bandwidth more effectively, since the number of iterations required to find a solution is less than for the resource-oriented case. However, if a solution is not found within the appropriate time limit, which is very probable at high load levels, then a feasible solution needs to be found first, which can mean a less-than optimal allocation of resources. The price-oriented algorithm also suffers from the requirement of a tuning phase due to the complexity of the utility model.

The use of a market-based multi-agent system can prove a valuable tool in the field of distributed IN load control. There are still open issues related to the algorithms presented here that need to be investigated: for example, the scalability of the algorithms with respect to the size of the network. In addition, how the algorithms can be adapted to a more complex IN architecture, for example, one in which multiple network operators co-exist.

Acknowledgements

This work was partially supported by the EU-funded ACTS project AC333 MARINER. The authors wish to acknowledge the valuable contribution that their colleagues in MARINER have made to this chapter.

19.7   References

Arvidsson, A., Jennings, B. and Angelin, L. (1999) On the Use of Agent Technology for Load Control, in *Proceedings of 16th International Teletraffic Congress.*

Atai, A.H. and Northcote, B.S. (1996) AIN Focused Overloads – A Review of USA CCS Network Failures and Lessons Learned, in *Proceedings of ITC Mini-Seminar on Engineering and Congestion Control in Intelligent Networks*, Melbourne, Australia.

Bedoy, J. et al. (1998) Study of Control Mechanisms for the Effective Delivery of IN Services, in *Proceedings of the IEEE International Communication Conference (ICC '98)*, pp. 1738–1742.

Bikhchandani, S. and Maner, J.W. (1997) Competitive Equilibrium in an Exchange Economy with Indivisibilities, *Journal of Economic Theory*, **74**, 385–413.

Clearwater, S.H. (Ed.) (1996) *Market-Based Control: A Paradigm for Distributed Resource Allocation*, World Scientific Press.

Friedman, D. and Rust, J. (Eds.) (1993) *Double Auction Market*. Perseus Books.

Hac, A. and Gao, L. (1998) Congestion Control in Intelligent Network, in *Proceedings of the IEEE International Performance, Computing & Communications Conference*, pp. 279–283.

Kawamura, H. and Sano, E. (1996) A Congestion Control System for an Advanced Intelligent Network, in *Proceedings of NOMS '96 - IEEE Network Operations and Management Symposium*, pp. 628–631, Kyoto, Japan.

Kihl, M. and Nyberg, C. (1997) Investigation of Overload Control Algorithms for SCPs in the Intelligent Network. *IEE Proceedings-Communications*, **144(6)**, 419–424.

Lodge, F., Botvich, D. and Curran, T. (1998) A Fair Algorithm for Throttling Combined IN and Non-IN Traffic at the SSP of the Intelligent Network, in *Proceedings of Intelligence in Service & Networks (IS&N '98)*, Antwerp, Belgium.

MARINER (1998) MARINER ACTS Project. Web site: http://www.teltec.dcu.ie/mariner/ [14 June 2000].

Pham, X.H. and Betts, R. (1994) Congestion Control in Intelligent Networks. *Computer Networks and ISDN Systems*, **26(5)**, 511–524.

Sidi, M., Lui, W.Z., Cidon, L. and Gopal, I. (1993) Congestion Control through Input Rate Selection. *IEEE Transaction on Communications*, **41(3)**, 471–477.

Smith, D.E. (1995) Ensuring Robust Call Throughput and Fairness for SCP Overload Controls. *IEEE/ACM Transactions on Networking*, **3(5)**, 538–548.

Smith, D.E. and Northcote, B. (1998) Service Control Point Overload Rules to Protect Intelligent Network Services. *IEEE/ACM Transactions on Networking*, **6(1)**, 71–81.

Wellman, M.P. (1993) A Market-Oriented Programming Environment and Its Application to Distributed Multicommodity Flow Problems. *Journal of Artificial Intelligence Research*, **1(1)**, 1–22.

Wellman, M.P., Walsh, W.E., Wurman, P.R. and MacKie-Mason, J.K. (1998) Auction Protocols for Decentralized Scheduling. *Games and Economic Behaviour*.

Yamaki, H., Wellman, M.P. and Ishida, T. (1996) A Market-Based Approach to Allocating QoS for Multimedia Applications, in *Proceedings of the 2nd International Conference on Multi-Agent Systems (ICMAS-96)*, pp. 385–392, Kyoto, Japan.

Ygge, F. and Akkermans, H. (1997) Duality in Multi-Commodity Market Computations, in C. Zhang and D. Lukose (Eds.) *Proceedings of the Third Australian Workshop on Distributed Artificial Intelligence*, pp. 65–78, Perth, Australia.

20

Victor – Proactive Fault Tracking and Resolution in Broadband Networks Using Collaborative Intelligent Agents

J. Odubiyi, G. Bayless, E. Ruberton

20.1 Introduction[1]

Tracking end-to-end Virtual Connections (VCs) to locate the sources of VC and trunk performance degradations on Concert's global broadband (ATM, frame and IP) networks is currently performed using some labour-intensive methods, according to Concert network engineers[2]. To determine the health status of each VC, network support engineers typically execute several commands to identify the path from the entry node to the exit node for each subnet. This requires the coordination of several people who must independently review separate databases to check information on network elements and communicate IP addresses of network elements over the telephone. This process is both error-prone and time-consuming.

Network fault management involves the physical monitoring of network elements, fault assessment and fault mitigation. Traditional vendor-supplied network-element management systems present a never-ending stream of alarm information to the operator consoles in real time. Depending on the state of the network, the volume of alarms can range from tens to hundreds of messages per minute. Once an alarm is reported, or a trouble ticket is opened, network operators at the Network Management Centres perform problem diagnosis to ascertain the problem and its exact location. During this manually intensive and time-

[1] This research was funded by BT Corporate Research and Technology Programme Office at Adalstral Park, Martlesham Heath, Ipswich, UK.

[2] See Acknowledgements.

consuming process, network performance often continues to deteriorate. Network fault assessment in a distributed environment is essential for prompt diagnosis and problem resolution to provide quality of service to customers.

Victor provides an intelligent approach to automatic fault assessment once a trouble ticket is generated. The Victor prototype automatically checks for trouble tickets, performs fault assessment and reports the fault to the user. In addition, it supports proactive fault management by regularly polling for specific performance criteria such as the networks' trunks and port statistics. It also calculates performance trends and alerts the operator to potential faults in the network. A rapid implementation of mitigation strategies may prevent a network anomaly from escalating into an unacceptable level of performance, which is a possible occurrence with the existing tool set.

A multi-agent system (MAS) employs a multi-layered system architecture based on the roles performed by each agent. Hayzelden and Bigham (1998) applied the multi-layered approach for ATM virtual path resource management. We applied it in Proteus for ATM network performance management (Odubiyi, Meekins, Huang and Tracy, 1999), and in SAIRE, an agent-based search engine (Odubiyi et al., 1997). The role modelling strategy has also been implemented successfully for generic partial goal planning (Lesser, 1998), and for business process management (Wooldridge, Jennings and Kinney, 1999).

The Zeus collaborative agent-building tool kit (Ndumu, Nwana, Lee and Collis, 1999) supports the role modelling process. Therefore, the challenge that we face in this project is not how to build an operational MAS, but how to build an open MAS (Steels, 1998), where the agents can adapt to changing operational environments. Since the agents are autonomous (Heckman and Wobbrock, 1998; Dyer, 1999), control strategies used to manage conventional software programs are not suitable. Due lack of space, we cannot present a detailed implementation of our approach in this chapter.

Victor relies on distributed and collaborative intelligent processing agents to associate, correlate and combine knowledge and information from multiple operations databases. The distributed sources include information about the Cisco[3] switches, concentrators and cards stored in network operations' databases, trouble tickets stored in a Trouble Management System (TMS) database, VC configuration parameters stored in data warehouses, and network trunk utilisation statistics stored in the network Management Information Base (MIB). This effort builds on our previous research prototype, Proteus (Odubiyi et al., 1999), a multi-agent system with machine learning capability that predicts ATM link bandwidth utilisation.

The remainder of this chapter provides a brief background on Concert's global network, followed by our Victor design strategy and implementation architecture. We present the development environment and the user interface, followed by our future plans for the project and concluding remarks.

20.2   Background: Concert Global Managed Platform

The Concert Global Managed Platform (GMP) is a global network consisting of a number of Global Points of Presence interconnected by International Private Line Circuits (IPLCs).

---

[3] Cisco is a trademark of Cisco Systems.

These IPLCs are leased from various carriers around the world. They are T1/T3 and E1/E3 circuits. Concert installs T3 and E3 circuits where tariffs and traffic levels justify this higher core bandwidth, and where they are available from the Public Telephone and Telegraphs. The IPLCs interconnect NET IDNX 90 bandwidth managers at the edge subnets. These are used to provide a managed transmission network, which supports the Concert service platforms. The service platforms include X.25 switches that support Concert Packet Services, frame relay switches that support Concert Frame Relay Service, ATM switches for Concert ATM services, and voice switches that support Concert Virtual Network Services. By providing a common transmission network to support its service platforms, Concert can pool the bandwidth requirements of these platforms. This allows Concert to gain economies of scale in procuring capacity from third-party carriers, and allows it to quickly provide bandwidth to the service platforms when required.

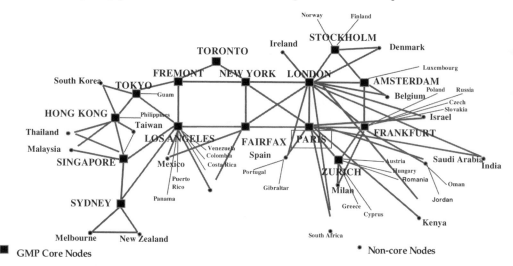

Figure 20.1   Concert Global Managed Platform

The geographic coverage and bandwidth size of the GMP grows monthly, so Figure 20.1 depicts a snapshot that may not represent a true picture of the system. The GMP has about 1000 circuits in 52 countries with over 2300 Mbits of shared capacity. At the time of writing, Concert provides six service configurations to its customers, based on their existing networks and requirements. These services include: (i) ATM service using Network-to-Network Interfaces (NNIs); (ii) ATM/FR interworking in the core subnet; (iii) ATM/FR interworking in an edge subnet; (iv) frame relay using FR NNIs; (v) frame relay NNI on one side of the core and ATM NNI on the other; and (vi) frame relay in the core with two ATM NNIs at the edges. The GMP also includes IP service configurations.

20.3   Assumptions

To aid the reader in understanding the limitations of the current version of Victor, this section describes the types of network data used in running the prototype. The Victor implementation is driven by near real-time performance statistics on network elements (ATM and frame relay) from Concert's hourly data dumps provided by a data warehouse. The data are collected on pre-specified collection intervals. The network topology data are obtained from an *Informix* network operations database. Criteria for performance are limited to the investigation of:

- Trunk loading (e.g. cells/frames/packets)
- Routing for VCs
- NNI selection
- Network parameters (e.g. port statistics and traffic types)
- Connection parameters (e.g. cells/frames/packets received and transmitted)
- Customer adherence to subscribed committed information rates.

20.4   Functional Objectives of Victor

The Victor demonstration system is being developed to achieve the following objectives:

1. Display the core network topology and indicate problem network elements through an effective user interface with colour codes.
2. Implement collaborative intelligent agents to detect the location of faults on specific VCs for which trouble tickets have been raised.
3. Summarise fault assessment information to provide network operators and customers with decision support regarding the health status of VCs and associated switches, cards, trunks, and NNIs, both on a continuous basis and on demand.
4. Summarise fault prediction information to anticipate the performance degradation of network elements for various VCs on a continuous basis.
5. Adapt existing trending algorithms to trend network element performance statistics and support proactive fault management.
6. Support attainment of network performance goals (i.e. 99.9% availability) by continuously monitoring the network elements and updating the Fault Table with critical information to enhance intelligent agent actions and advice.

20.5   Design and Implementation Strategy for Victor MAS

The core of the Victor system is the agent layer. The agent layer consists of a number of agents that provide the control mechanism for achieving the desired objectives of the Victor system. Figure 20.2 illustrates the design architecture of the Victor system prototype. The agent layer has a hierarchical structure consisting of groups of agents with different high-level roles or sublayers: user interface; agent services management; problem solving; network monitoring and performance trending; and interfaces with heritage-network operations data sources. The agent services and management sublayer has only a single agent, the Agent Services Manager (ASM), since the number of agents required for the prototype is small. This layer also provides coordination capability for supporting directory services, 'yellow pages', and life-cycle servicing tasks for the agents. To support any demand for additional agents, peer ASMs can be implemented to manage distributed agent clusters. Interactions among the agents from different clusters can be achieved via peer-to-peer agent communications.

The agents in the multi-agent system perform various functions, often in collaboration, to obtain data on network performance statistics residing in the network's management information base. A trouble-ticket management system provides information on existing faults on network elements and network topology from the network operations database. We are working on each of the six functional objectives of Victor in parallel; each is at a different stage of completion.

One of the specific roles of the network element monitoring and trending agents is that of determining if there is a fault or if there is likely to be a fault (current and predicted fault information). This is performed by using the following procedure:

- Check if VC routing is reasonable.
- Check NNI utilisation for acceptability criteria.
- Check Trunk utilisation for acceptability and error criteria.
- Check VC queuing and discards using VC interval statistics.
- Check Trunk queuing and discards using Trunk interval statistics.
- Check Port queuing and discards using Port interval statistics.
- Check VC utilisation.

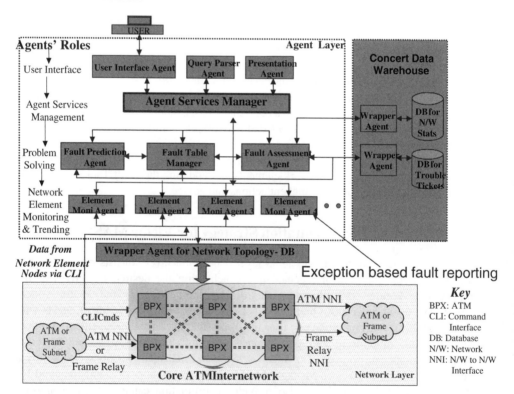

Figure 20.2   The Victor system implementation architecture

The network element monitoring agents employ an exception-based fault reporting process. This process implies that each agent does not report a fault unless it violates a specified threshold value. The strategy helps in reducing the volume of message traffic between the agents and hence network traffic. In addition to monitoring the network elements for existing faults, the network element monitoring and trending agents employ three locally weighted time series algorithms (Atkenson, Moore and Schaal, 1996) to trend the network performance statistics for use by the fault prediction agent.

The user interface agent accepts and converts user requests to a FIPA[4] compliant-Agent Communication Language (ACL), and presents summary results back to the user. It also receives information from the presentation agent to show the fault assessment or prediction results to the network operator or customer.

The fault prediction and the fault assessment agents are the problem solving agents. The fault-prediction agent assists in proactive fault management of the network. It has the capability to use historical as well as current fault information, within respective network elements, to predict potential problems along the path of specified VCs. The fault assessment agent provides reactive fault tracking activity for the VCs.

The wrapper agents provide interfacing functionality between the information sources and the agent layer. The wrappers retrieve operations data using Java Data Base Connectivity (JDBC) protocols and deliver them to the agents using the ACL format. Hourly transport of data from the databases helps reduce the traffic load on the network. Daily data collections are played back to simulate real-time data feeds into Victor agents.

20.6   Victor Development Environment

The agents in Victor system prototype run on distributed computer platforms including Sun SPARC Stations running the Solaris operating system and on Pentium-class PCs running the Windows-NT and Linux operating systems. Some of the agents were developed with the Java Expert System Shell (JESS), and the database wrapper agents were developed with Zeus, a Java-based agent-development tool kit. Both shells embed CLIPS (C-Language Interface Production System) for inferencing mechanisms. All the agents communicate in a FIPA-compliant Agent Communication Language (ACL) thus eliminating any drawbacks of heterogeneous hardware and software development environments. The wrapper agents employ JDBC/ODBC to retrieve network performance and topology data from distributed operations databases.

20.7   Victor User Interface

A sample Web page of the Victor user interface is depicted in Figure 20.3. The interface permits the user to view the network topology and drill down to view the cards, ports and connections for any switch or other network elements.

---

[4] Foundation for Intelligent Physical Agents. Web site: http://www.fipa.org [18 May 2000].

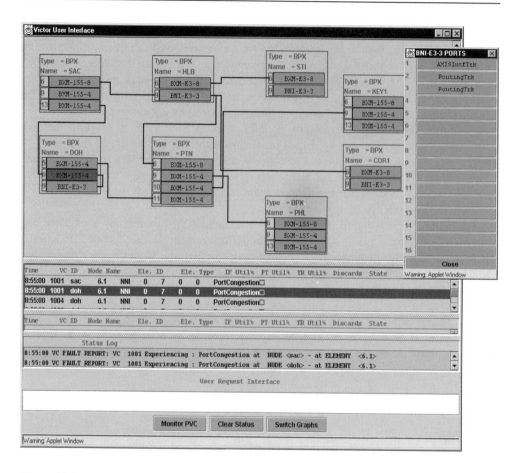

Figure 20.3 The Victor graphical user interface

The middle segment of the user interface displays specific faults or levels of performance degradation for affected network elements, and the impacted customer virtual connections. A user request input panel is provided at the bottom part of the interface to allow user input of semi-structured requests on the health status of specific VCs.

20.8   Future Plans

We plan to increase the number of agents in Victor to support all the network elements in the Global Managed Platform. This demands the development of strategies that make the agents perform in the hard real-time environments typically encountered in this domain. To survive in such environments the agents need to be adaptive with repeatable and assured behaviour to avoid undesirable consequences. The goal-based, self-monitoring methods employed by the Automated Reasoning group at Honeywell Technology Center and at the University of Michigan (Musliner et al., 1999) in developing a Self-Adaptive Cooperative Intelligent Real-time Control Architecture (SA-CIRCA) will be investigated to guide future development of Victor. We will also incorporate survivability strategies for broadband networks proposed by David Johnson at BT Laboratories (Johnson, 1996). The survivability methods employed include preplanning tasks performed by the agents and allow the agents to be bounded by available computational resources and real-time demands – bounded rationality.

20.9   Concluding Remarks

We have presented Victor, a multi-agent system that proactively and reactively tracks faults affecting network elements along the path of customers' virtual connections in broadband networks. The agents collaborate in processing and presenting information on the health and status of the VCs to the users. Due to the dynamic nature of the operations of the domain, successful strategies for managing hard real-time environments are being employed to guarantee desirable behaviour of the agents and avoid disaster in the operations of the telecommunication network. The system is being developed with the goal of avoiding any agent's actions that will negatively impact the minimum goal of 99.9% system availability. The agents' inefficient use of the network resources would negatively impact the achievement of the goal and plant negative opinions on the efficacy of agent-based systems.

The need to work closely with network experts to obtain a sound understanding of the domain is critical to any successful development and deployment of agent-based systems. Working closely with network domain engineers, we developed our first prototype with simulated data. That experience exposed us to some limitations of simulated data, but it gave us a better understanding of several operational scenarios. In our follow-on task, we used the data from a network laboratory, which gave us direct exposure to the operations of network elements (switches, trunks, ports, etc.). The experience was helpful, but because the laboratory environment supports several users the networks are reconfigured periodically to meet the needs of other users. Finally, we requested and obtained permission to use realistic network data from various sources. Researchers should not underestimate the level of effort required to acquire realistic network operations data – it is possible to spend an inordinate amount of time locating pertinent data. To succeed, a researcher must have interested network operations personnel and engineers in his or her corner.

The network operations domain is a natural environment for deploying multi-agent systems because the system, the experts and the users are highly distributed, demanding sophisticated coordination and control strategies to harness desirable behaviour of the agents in real time.

Acknowledgements

We are grateful for the support of Dave Clark, Jay Witek, Jim Moore and Guy Hanacek at Concert Network Engineering group in Reston, Virginia, and John Haynes of Concert in London, England; they provided substantial expertise in helping us formulate the problem and identifying the appropriate network data to support this research. We are also thankful to Ray Godwin and Rod Benton at Concert Global Network Management Center in Atlanta for providing data acquisition support, and to Professor George Tecuci of George Mason University, Fairfax, Virginia, USA and Marcus Thint for their helpful contributions to the project and this chapter.

20.10   References

Atkenson, C.G., Moore, A.W. and Schaal, S. (1996) Locally Weighted Learning. Technical Report. Georgia Institute of Technology. Available: http://www.cc.gatech.edu/fac [15 June 2000].

Dyer, D.E. (1999) Multiagent Systems and DARPA. *Communications of the ACM*, **42(3),** 53.

Hayzelden, A.L.G. and Bigham, J. (1998) A Heterogeneous Multi-Agent Architecture for ATM Virtual Path Network Resource Configuration, in *Proceedings of Intelligent Agents for Telecommunications Applications*, pp. 45–59, Springer-Verlag.

Heckman, C. and Wobbrock, J.O. (1998) Liability for Autonomous Agent Design, in K.P. Sycara and M. Wooldridge (Eds.) *Proceedings of the 2nd International Conference on Autonomous Agents (AGENTS-98)*, pp. 392–399, Minneapolis, MN.

Johnson, D. (1996) Survivability Strategies for Broadband Networks. BT Laboratories, B61, Martlesham Heath, Ipswich IP5 7RE, UK.

Lesser, V.R. (1998) Reflections on the Nature of Multi-Agent Coordination and Its Implications for an Agent Architecture. *Autonomous Agents and Multi-Agent Systems*. **1,** 89–111.

Musliner, D.J., Goldman, R.P., et al. (1999) Self Adaptive Software for Hard Real-Time Environments. *IEEE Intelligent Systems & Their Applications*. July/August, 23–29.

Ndumu, D.T., Nwana, H.S., Lee, L.C. and Collis, J.C. (1999) Zeus: A Toolkit and Approach for Building Distributed Multi-Agent Systems, in *Proceedings of the Third Annual Conference on Autonomous Agents*, pp. 360–361, Seattle, WA.

Odubiyi, J.B., Kocur, D. J., Weinstein, S.M., Wakim, N., et al. (1997) SAIRE - A Scalable Agent-Based Information Retrieval Engine, in *Proceedings of the First International Conference on Autonomous Agents,* pp. 292–299, Marina del Rey, CA.

Odubiyi, J.B., Meekins, G., Huang, S. and Tracy, Y. (1999) Proteus – Adaptive Polling System for Proactive Management of ATM Networks Using Collaborative Intelligent Agents, in *Proceedings of the Third Annual Conference on Autonomous Agents*, pp. 402–403, Seattle, WA.

Steels, L. (1998) The Origins of Syntax in Visually Grounded Robotic Agents. *Artificial Intelligence*, **103,** 133–156.

Wooldridge, M., Jennings, N.R. and Kinny, D. (1999) A Methodology for Agent-Oriented Analysis and Design, in *Proceedings of the Third Annual Conference on Autonomous Agents*, pp. 69–83, Seattle, WA.

21

Efficient Means of Resource Discovery Using Agents

S. Sugawara, K. Yamaoka, Y. Sakai

21.1   Introduction

Recently, large-scale networks such as ISDN and the Internet have become very popular. With the rising number of network users, the quantity of information flowing through the network has increased dramatically. Moreover, the volume and variety of information handled by network users are growing. However, because there is so much information on the network that is irrelevant to the user, they have to spend a lot of time searching through this information for the information they want. This problem is called 'resource discovery problem', and has become an important area of study.

In this study, we describe efficient methods of searching with agents for information on large-scale networks. We first define communication/processing cost and time cost as well as the information search problem. Then we classify the types of agents by their number and mobility. Finally, efficient methods are described for two types of agents.

21.2   An Outline of an Agent

The agents in this study perform the function of information searching. They also have autonomy, so when they search receive requests for required information (from now on, we call required information or the information that is being searched the 'target') from a user, they search for the target automatically. At the end of a search, the agent sends the search result (information about the target or a failure notice), and ends the search process.

21.3   Search Problem

Let us assume that there is a network consisting of many nodes and links, and many sources of information are located at the nodes. We call the nodes that have sources of information 'search nodes', and the node where the agent is located the 'base node'. An agent is located at the base node at the beginning of a search, and from there it starts to make inquiries to search nodes to find the target.

21.3.1   Communication and processing cost

The communication cost of a search is the cost of exchanging data between nodes where the agent or user is located and search nodes. The processing cost is the cost of determining whether the information located at a search node is the target. The actual definition of communication/processing cost depends on whether the agent is fixed or mobile.

21.3.2   Time cost

Time cost is the time required to determine whether the target exists at a certain search node. The actual definition of time cost also depends on the agent model.

21.3.3   Link capacity of the network

In this study, we assume that link capacity can be estimated and its time fluctuation during the search is very small.

21.3.4   Probability of target existence

In this chapter, the probability of target existence is assumed to be already known.

21.3.5   Search problem

Based on the assumptions mentioned up to this point, we define the search problem as a problem of 'obtaining a search method that minimises the sum of average communication/processing cost and average time cost and that uses agents'.

21.4   Models and Classification

In this study, we classify the agent models according to the number and mobility of the agents (Table 21.1).

The models of agents are based on the following:

1. One agent fixed at a base node accesses search nodes to try to find a target (single-fixed type).
2. One agent moves around search nodes to try to find a target (single-mobile type).
3. A number of agents fixed at base nodes access to search nodes separately to try to find a target (multiple-fixed type).

4. A number of agents separately move around search nodes to try to find a target (multiple-mobile type).

Table 21.1   Classification of agents. Reproduced from Sugawara et al (1999), (©1999 IEEE) and Sugawara, Yamaoka and Sakai (1999), (©1999 IEEE).

	Fixed	Mobile
Single	(1)	(2)
Multiple	(3)	(4)

In the remaining sections, we discuss models (1), (2) and (3), discuss search algorithms used in the models and examine their efficiency.

21.5   Single Fixed Model

21.5.1   Search problem

- For the case of a single agent, let us assume that
- N_d is the number of search nodes
- d_i is the node searched on the i th turn
- p_i is the probability of target existence on the search node d_i
- c_i is the communication/processing cost of searching the search node d_i
- T_i is the time cost of searching the search node d_i

The total search cost can be estimated as follows.
First, the total mean value of the communication/processing cost, E_i can be written as

$$E_c = c_1 + \sum_{l=2}^{N_d} \left\{ c_l \times \prod_{k=1}^{l-1} (1 - p_k) \right\}$$ (21.1)

Second, the total mean value of the time cost, E_t, can be described as,

$$E_t = T_1 + \sum_{l=2}^{N_d} \left\{ T_l \times \prod_{k=1}^{l-1} (1 - p_k) \right\}$$ (21.2)

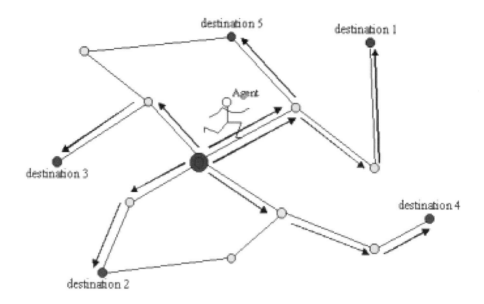

Figure 21.1   Searching by single-fixed agent. Reproduced from Sugawara et al (1999),
(©1999 IEEE) and Sugawara, Yamaoka and Sakai (1999), (©1999 IEEE).

The simple sum of the total mean value of communication/processing cost and time cost
can be considered as the estimated value of total search cost.

But in this chapter, we assume the estimated value of total search cost for evaluation
of search efficiency as follows.

$$E = W_1 \times E_c + W_2 \cdot E_t \tag{21.3}$$

Weights W_1 and W_2 (W_1, $W_2 \geq 0$, $W_1 + W_2 = 1$) can be decided according to the user's
preference. This makes the searching policy tunable to the user's needs, so that the user can
conduct time-priority searches or resource-priority searches.

21.5.2   Optimum search method

It is possible to show that calculating $p_j / (W_1 \times c_j + W_2 \times T_j)$ for each search node d_j,
and searching the nodes in descending order of the value is the optimum method. This
gives the lowest average of estimated value E. (The proof is omitted.)

The search which minimises $p_j / (W_1 \times c_j + W_2 \times T_j)$ can be done with minimum time
cost by setting the weights $W_1 = 0$, $W_2 = 1$. In this case, the search is conducted in

descending order of p_j / T_j. By setting the weights $W_1 = 1$, $W_2 = 0$, the search can be conducted with minimum communication/processing cost in descending order of p_j / c_j.

21.6   Single Mobile Model

21.6.1   Search problem

The estimated value E mentioned in Section V can be obtained as follows. First, the following values are assumed.

D is a set of numbers assigned to the base node and searched nodes (the assigned number of the base node is 0)

$\widetilde{c}_{i,j}$ is the communication/processing cost for moving between d_i and d_j ($i, j \in D$),

$\widetilde{T}_{i,j}$ is the time cost for moving between d_i and d_j ($i, j \in D$).

The mean communication/processing cost and mean time cost while the agent moves around the search nodes in the order of $d_1, d_2, \cdots, d_n$ can be expressed as

$$E_c = \widetilde{c}_{0,1} + \sum_{k=1}^{N_d - 1} \left\{ \widetilde{c}_{k,k+1} \times \prod_{l=1}^{k} (1 - p_l) \right\} \tag{21.4}$$

$$E_t = \widetilde{T}_{0,1} + \sum_{k=1}^{N_d - 1} \left\{ \widetilde{T}_{k,k+1} \times \prod_{l=1}^{k} (1 - p_l) \right\} \tag{21.5}$$

Then,

$$E = W_1 \times E_c + W_2 \times E_t \tag{21.6}$$

Accordingly, the problem to deal with here is to obtain the route around the search nodes in the network that minimises E. Finding the optimum solution is very difficult. For example, in the case of $W_1 = 1$, $W_2 = 0$ (that is, considering only communication/ processing cost), the problem is a kind of extension of the travelling salesman problem. Thus it is difficult to find an optimum way of searching within polynomial time.

In this chapter, we propose using an approximation algorithm that makes a small loop route for the search and inserts search nodes into the loop to cover all search nodes in the network.

21.6.2   Evaluation of proposed algorithm

The effectiveness of the proposed algorithm can be shown by simulations. Let us assume a 100-node lattice network in a 10 by 10 pattern. Search nodes are selected at random from these 100 nodes. We conducted three simulations for 10, 30, and 50 nodes. The time cost for moving from the base node or a search node to another search node is assumed to be constant ($\tilde{T}_{i,j}$ = 400). The communication/processing cost of moving along one link and calculating for target existence is assumed to be based on a uniform distribution from 80 to 120 (average 100). The probability of the target's existence at each node p_j is assumed to be based on the following discrete probability distribution (Figure 21.2).

$$p_j = \frac{1}{j \times \sum_{k=1}^{100} \frac{1}{k}} \tag{21.7}$$

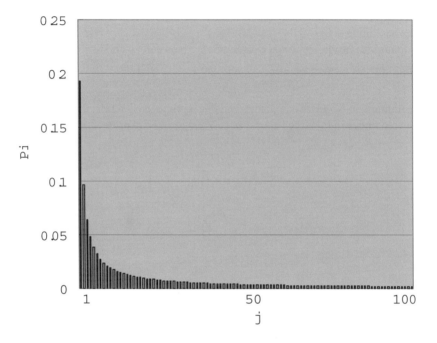

Figure 21.2 Discrete probability distribution. Reproduced from Sugawara et al (1999), (©1999 IEEE) and Sugawara, Yamaoka and Sakai (1999), (©1999 IEEE).

The probability is assumed not to change until the end of the search, and the initial location of the agent (or base node) is fixed at the centre of the lattice.

We show the simulation results for the two cases (W_1 = 1.0, W_2 = 0 and W_1 = 0.5, W_2 = 0.5), calculated under the assumptions and by the algorithm described above (see figures 21.3, 21.4). Each simulation was run 1000 times to obtain an average of the mean total costs. To evaluate the effectiveness of our algorithm, we conducted simulations using two other algorithms as follows:

- Probability order search (algorithm for comparison 1)
- Exhaustive search for optimum solution (algorithm for comparison 2)

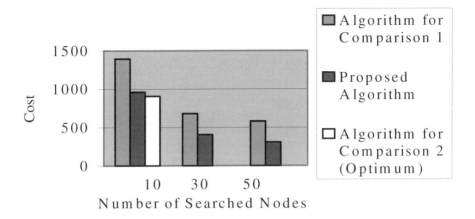

Figure 21.3   Total cost of single-mobile type (W_1=1.0, W_2=0). Reproduced from Sugawara et al (1999), (©1999 IEEE) and Sugawara, Yamaoka and Sakai (1999), (©1999 IEEE).

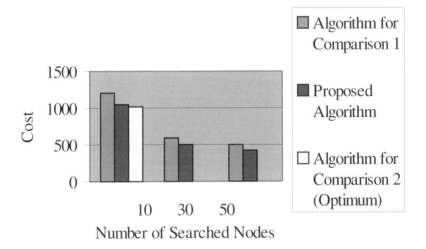

Figure21.4   Total cost of single-mobile type (W_1=0.5, W_2=0.5). Reproduced from Sugawara et al (1999), (©1999 IEEE) and Sugawara, Yamaoka and Sakai (1999), (©1999 IEEE).

The algorithm for comparison 2 calculates E for all permutations of search nodes, and takes the permutation giving the smallest E as the optimum solution. However, because this algorithm cannot calculate within polynomial time, we were only able to conduct a simulation with 10 search nodes. Despite the number of search nodes, the estimated value of mean total searching cost E is smaller for our algorithm than for the algorithm for comparison 1. In the case of 10 search nodes, the performance of our algorithm is nearly equal to that of the algorithm for comparison 2 which gives optimum solution.

In all algorithms, the greater the number of nodes to be searched, the smaller the value of E is. This is due to the increase in the number of nodes that have a high probability of target existence. Because the values of probability are randomly assigned to each node from a distribution like the one in Figure 21.2, the number of nodes with a high probability increases as the number of search nodes increases.

21.7   Multiple Fixed Model

21.7.1   Search problem

In the case of searching using plural fixed agents, each agent inquires searched nodes of which it is in charge.

Practical searching mechanism is as follows. The plural agents continue to search until they find a target or finish inquiring all searched nodes of which they severally are in charge. When one of the agents find a target, the agent sends a message to other ones to let them quit searching.

Here, we assume that the cost to communicate among agents is small enough to be neglected.

Under the assumptions mentioned above, the problem to discuss is as follows. 'Obtain the searching policy (the assignment of searched nodes to agents and the searching order of searched nodes for each agent), which makes weighted sum of communication/processing cost and time cost the smallest, in the case that the number of agents and the locations of base nodes are given'. In this chapter, the number of agents and the locations of base nodes are given, and the agent allocation problem is not covered.

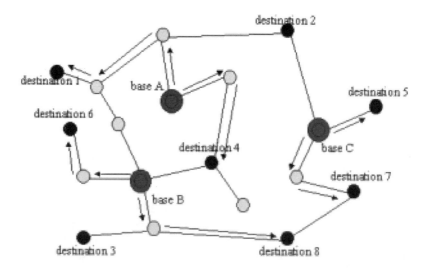

Figure 21.5 Searching by multi-fixed agents. Reproduced from Sugawara et al (1999).

21.7.2   Proposed algorithm

We propose two algorithms below, which is an extension of the searching algorithm for single-fixed type:

- Basic algorithm
- Modified algorithm

At first, we state the basic algorithm. Putting communication/processing cost and time cost for searching between an agent x and a searched node y equal to $c_{x,y}$ and $T_{x,y}$, this algorithm distributes the same number of searched nodes among agents in descending order of $p_y / (W_1 \times c_{x,y} + W_2 \times T_{x,y})$.

Then, we propose a modified algorithm based on the basic algorithm. This algorithm in practice consists of tree processes, that is, Searched Nodes Distribution, Initial Scheduling, and Dynamic Modification.

(i) Searched Nodes Distribution:
In the case of $W_1 \geq W_2$, each searched node is assigned to the agent which can search in the smallest communication/processing cost.

In the case of $W_1 < W_2$, each searched node is assigned one by one to each agent in descent order of target's existence probability, and finally all searched nodes are distributed to agents. But in the case of assignment, trying to make the amount of time cost for searching the same in all agents, each searched node is assigned to the agent which requires as small a communication/processing cost as possible.

As a result, in the case that communication /processing cost is important, this algorithm averages the cost among all agents, and allows the least time cost. On the other hand, in the case that time cost is important, this algorithm averages the cost and allows the least of communication/processing cost.

(ii) Initial Scheduling:
Each agent i basically searches the assigned searched nodes j in descent order of $p_j / (W_1 \times c_{i,j} + W_2 \times T_{i,j})$. This process continues until one of agents finds a target or all the searched nodes of which it is in charge are searched.

(iii) Dynamic Modification:
Only in the case mentioned below, is Initial Schedule changed. Each agent i calculates $p_j / (W_1 \times c_{i,j} + W_2 \times T_{i,j})$ and selects the node which gives the largest value of it among all searched nodes j which have not been searched yet, every time after the failure of target discovery.

When the selected node is not the one which is planned to be searched by the agent next in Initial Schedule, the expected value of total cost is calculated in two cases, the Initial Schedule and a changed one, and if the changed one is advantageous, Initial Schedule is changed.

The agent i which has lost the node to search for that change, selects the searched the node j which gives next largest value of $p_j / (W_1 \times c_{i,j} + W_2 \times T_{i,j})$ to search.

(iv) Each agent finishes its task and extingguishes itself when it has finished searching all nodes which are assigned to it.

(v) When an agent has found a target or all agents have finished their task, this searching ends.

Here, we mention the time complexity of the two algorithms above. Assuming that the number of searched node is $O(D)$, the time complexity of the two algorithms is $O(N(D)^2)$. Therefore, these algorithms end their process within polynomial time, and are in fact useful.

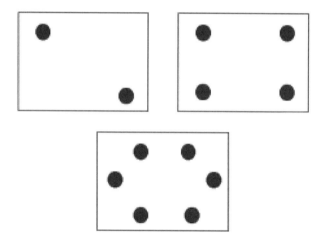

Figure 21.6 Location of the agents. Reproduced from Sugawara et al (1999), (©1999 IEEE).

21.7.3   Evaluation of proposed algorithm

In a similar way to the former section, we assume a lattice network which consists of 10 by 10 nodes. But every time cost for 1 search for a searched node by an agent is set at constant. And its value $T_{i,j}$ equals 1600 in order to balance with communication/ processing cost.

The number of agents is set in three ways, that is, at 2, 4, and 6 on the lattice like the Figure 21.6.

Under the conditions above, we show the samples of simulation results in the two cases of $W_1 = 1$, $W_1 = 0$, and $W_1 = 0.5$, $W_1 = 0.5$ (Figure, 21.7, 21.8). Those simulations give the mean total costs of 1000 calculation times by two algorithms.

In Both cases of $W_1 = 1$, $W_1 = 0$ (Figure 21.7) and $W_1 = 0.5$, $W_1 = 0.5$ (Figure 21.8), mean total cost of the modified algorithm is smaller than that of the basic algorithm. And the more agents are used, the better the effect of the modified algorithm works. This result shows as follows. When the number of agents increases, it is clear that the number of searched nodes assigned to each agent in order to search is raised. Then it is inappropriate to assign the same number of searched nodes forcibly to each agent.

Although the number of agents increases, the mean total cost does not decrease. This shows that too many agents are useless in the case of having some searched nodes with sufficiently high probability of target's existence. In the case of no information of target's location, we confirmed that the more agents are used, the smaller mean total cost is required in both algorithms.

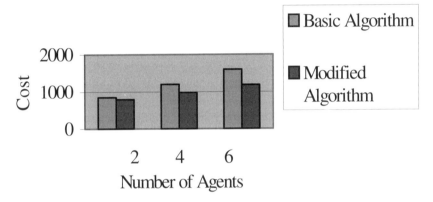

Figure 21.7 Total cost of multi-fixed type (W_1=1.0, W_2=0). Reproduced from Sugawara et al (1999), (©1999 IEEE).

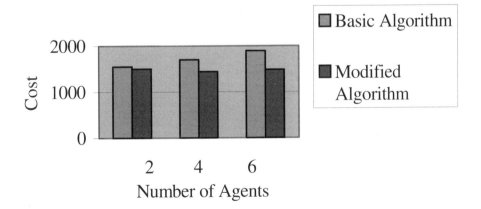

Figure 21.8 Total cost of multi-fixed type (W_1=0.5, W_2=0.5). Reproduced from Sugawara et al (1999), (©1999 IEEE).

21.8   Conclusion

In this chapter, we classified search problems with agents according to the number of agents and their mobility. We then proposed efficient search methods for a single fixed agent and a single mobile agent under the assumption that the probability of the target's existence is known. Furthermore, we showed the efficiency of these methods.

21.9   References and Further Reading

Bowman, C.M., Danzig, P.B., Manber, U. and Schwartz, M.F., (1994) Scalable Internet Resource Discovery: Research Problems and Approaches, *Communications of ACM*, **37(8)**, pp. 98-114.

Sakai, Y., Yamaoka, K. and Sugawara, S., (1995) A Study on Information Searching Method Based on Entropy, *IEICE Transactions*, **J78-A(7)**, pp. 894-896.

Sugawara, S., Ohshima, T., Yamaoka, T. and Sakai, Y. (1999) A Study on Efficient Information Search with Agents in Large-Scale Network, *The Journal of the Institute of Electronics, Information and Communication Engineers*, **J82-B(6)**. pp. 1115-1125.

Sugawara, S., Yamaoka, K., and Sakai, Y., (1997) A Study on Image Searching Method in Super Distributed Database, *IEEE GLOBECOM'97*.

Sugawara, S., Yamaoka, K., and Sakai, Y. (1999) A Study on Efficient Information Searches with Agents for Large-Scale Networks. *Proceedings of IEEE Globecom'99*, pp. 1954-1958.

22

Evolving Routing Algorithms with Genetic Programming

E. Lukschandl

22.1   Introduction

The purpose of a communication network is to fulfil demands between pairs of nodes, e.g. to transmit a package or to establish a connection for a certain amount of time between pairs of nodes. Because the fulfilment of such a demand uses one or more kinds of limited resources in the nodes and/or the links of the network, an increase in demand will sooner or later exhaust the resources, depending on the scheme of their allocation, i.e., depending on the decisions made about which path from a set of paths to choose when fulfilling each demand. This is called the routing problem (Bertsekas and Gallager, 1992; Steenstrup, 1995).

The focus of this chapter is the routing problem in circuit-switched communication networks, especially the study of how to reduce the number of telephone calls that cannot be successfully connected due to congestion in one or more parts of the network. It is well known that reducing this number, even by only a small percentage, can save large amounts of money. Because of this, there is considerable interest in this field and researchers have come up with a wide variety of solutions. Related work has mainly addressed one of two classes of problem related to resources:

1. Finite capacity and/or propagation delay in the links
2. Finite buffering and/or processing capacity in the nodes.

Examples for adaptive routing algorithms can be found in the following categories:

1. Ant algorithms, e.g., Schoonderwoerd, Holland, Bruten and Rothkrantz (1997)
2. Blocking island hierarchy routing, e.g. Frei and Faltings (1998)
3. Market-based algorithms, e.g. Gibney and Jennings (1998)
4. Fuzzy decision rules, e.g. Arnold, Hellendoorn, Seising, Thomas and Weitzel (1997)
5. Genetic programming, e.g. Lukschandl, Borgvall, Nohle, Nordahl and Nordin (1999).

22.2   The Model

One of the main tools used for this research is a simulator developed by the author. There are several reasons for using a simulator instead of, or in addition to, a real-world network, and the most important one is that a simulator allows us to study the effect of different routing schemes when using the same input. The input for the simulator is a *call file* containing a time-stamped number of calls with a number of attributes as will be described later. A separate program, the *call file generator*, creates these calls randomly with user-defined probabilities for the values of the attributes. Another reason for using a simulator is simply the difficulty in using real, live data. In addition, the fact that we do not have real logged data available makes the call file generator a necessity. Both the simulator and the call file generator are implemented in Java.

22.2.1   The topology

As a start, the northern part of BT's Synchronous Digital Hierarchy Network has been used. It consists of 13 nodes and 23 links. When the whole network has been modelled, we will be able to make comparisons with the work of Schoonderwoerd et al. (1997) and Appleby and Steward (1994) who used the same topology and made the same assumptions as defined below. The topology is defined by listing all pairs of directly connected nodes in a text file.

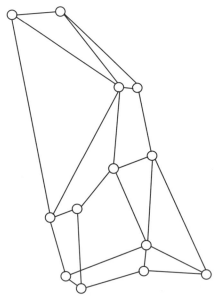

Figure 22.1   The northern part of BT's synchronous digital hierarchy network

22.2.2 Assumptions and modelled features

- The nodes contain resources of limited capacity
- Processing of data in the nodes takes zero time
- The links between the nodes have unlimited transportation capacity
- Data transmission along the links takes zero time.

22.2.3 The main data structures

The main data structures used are:

1. *Calls*: a call has the following attributes: source node id, destination node id, start time, duration, type, status.
2. *The Clock*: the clock is a simple counter defining the progress of time, which is thus discrete.
3. *The Event Queue*: the event queue is a time-sorted list of events. At this stage of the project these are only messages to agents. At a later stage other types of events will be introduced, like topology-changing events.
4. *Messages*: for the time being there are only two types of messages: *connect* and *disconnect*. Both messages have a call as an attribute. Other attributes are a time-stamp and receiver and sender addresses.
5. *Nodes*: nodes model telephone switches and hold certain state information, e.g. its load capacity, the current load on the node, a list of distances to all other nodes, a list of neighbour nodes, and a list of currently ongoing calls the paths of which contain the node.
6. Of central importance is the term *connection resistance* (or *c-resistance*) of a node. It is defined as an arithmetic function which takes as parameters a chosen number of state variables from the environment.
7. *Agents*: agents reside in nodes; they receive and send messages, and can read and update a subset of the state variables of the network, especially those of their own nodes, but also attributes of calls.
8. *The Network*: the network is primarily a list of nodes. The agents in the network form an *agent society*.

22.2.4 The dynamics of the model

The purpose of the network is to establish paths (connections) between pairs of nodes in order to enable telephone calls along these paths (routes).

A separate telephone call generator program creates a list of calls that are written into a call file, sorted by start time. Input to the call generator is the desired number of time steps of the simulation, and the values of a call's attributes are chosen randomly, with certain probabilities.

When the simulation is started the topology file is read into memory and the network created; the calls are then read from the call file into the event queue as connect messages.

Time is measured by incrementing the counter of the clock. The start time of the first call in the queue is compared with the time of the clock, and the clock is allowed to tick on until its time corresponds to the start time of the call. At that time the whole path from source to destination node is computed, i.e. time stands still until the call is connected from source to destination nodes or the connecting process has failed somewhere on the way.

When an agent at a node gets a connect message, the following rules apply:

1. If the node's capacity is exhausted, i.e. the current load equals the given maximum load, the connecting process failed, and the call's status is set to 'blocked'. The agent sends the message 'disconnect' to the agent that sent the connect message.
2. Else if the call has been handled by the node before, the connecting process failed, and the call's status is set to 'circular'. As above, the agent sends the message 'disconnect' to the agent who sent the connect message.
3. Else the agent increments the current load of the node, and:
   a)   If the node is identical with the destination of the call, the call connection is a success; the call duration is added to its start time, and a disconnect message is sorted into the message queue at the right place.
   b)   Else, if one of the neighbour nodes corresponds to the destination node of the call, the agent sends the message 'connect' to the neighbour agent.
   c)   Else, the agent uses its knowledge of the states of the neighbour nodes to compute their c-resistances, and sends a connect message to the one with the lowest c-resistance.

When an agent gets a disconnect message (it is always assumed that the disconnecting is initiated by the agent in the destination node), the following rule applies:

1. It decrements the current load of the node, and:
   a)   If its node is the source node of the call, the disconnecting process terminates.
   b)   Else, it sends the message 'disconnect' to the agent from which it received the connect message for this call.

22.3   Research Aims

The purpose of the agent society is to route a set of calls, distributed over a certain time interval, in such a way that the allocation of the resources is optimised, and we want to show one or more of the following:

1. The decision-making algorithm in the agents can automatically be developed by means of evolutionary programming techniques, especially by genetic programming.
2. Dynamic routing is superior to static routing.
3. Using local information only is superior to using global information.
4. Further, we want to investigate what kind of local knowledge is most important in order to make good decisions.

22.4   The Java Method Evolver (JME)

The Java Method Evolver is the other major tool for our research. It is a specialisation of the Java Bytecode Genetic Programming Engine (JBGP) developed by the author and others at EHPT Sweden AB (Lukschandl, Holmlund, Modén, Nordahl and Nordin, 1998a; Lukschandl, Holmlund, Modén, Nordahl and Nordin, 1998b).

John Koza, who published his original idea of genetic programming in 1992, writes in a later book: 'Genetic programming is an automatic programming technique for evolving computer programs that solve (or approximately solve) problems. Starting with a primordial ooze of thousands of randomly created computer programs composed of appropriate programmatic ingredients, a population of computer programs is progressively evolved over many generations using the Darwinian principle of survival of the fittest, a sexual recombination operation, and occasional mutation' (Koza, Goldberg, Fogel and Riolo, 1997). A detailed description of genetic programming is beyond the scope of this chapter, but a good overview can be found in Banzhaf, Nordin, Keller and Francone (1998).

The JME automatically evolves Java methods, i.e. functions. The specialisation is characterised by the fact that the method takes (any number of) real parameters, returns a real value, and uses only the four basic arithmetical operators: addition, subtraction, multiplication and division.

As described above the essence of genetic programming is the random creation of thousands of individuals, in our case methods represented as Java bytecode. In order for the JME to work, each method has to be evaluated. Evaluation is done by loading the method into the simulator and running the simulator for a certain period of time determined by the chosen call file. Depending on what is to be optimised, either the number of lost calls or the lost revenue is passed back to the JME as fitness of the individual, at the end of each run of the simulation.

22.5   The Experiments

When an agent has to decide to which neighbour node it should route a call, it invokes the c-resistance function in each neighbour node and picks the one returning the lowest value. Several experiments have been conducted. All had the following conditions in common:

1. The same network topology, with a load capacity of 8 in each node, is used.
2. The 250 calls are evenly distributed across 250 time-steps.
3. The 250 source destination pairs are drawn with equal probability.
4. The call duration is evenly distributed in the time interval [18, 22] time-steps.
5. Calls to a dedicated node – e.g. functioning as a gateway to the international network – have a value of 10 monetary units, whereas the others have a value of one unit.

During each run of the simulation different kinds of data are summarised, e.g.:

- The number of blocked calls
- The number of circular calls
- The value of the successfully routed calls.

22.5.1 Experiment A

The optimisation goal is to minimise the number of lost calls. We call this environment A of the agent society. The input for the c-resistance is:

- The distance from the node to the destination node, measured as the number of links; and
- The current load of the node, measured as percentage of the maximum load.

22.5.2 Experiment B1

In this case the optimisation goal is to maximise the total revenue of the calls. We call this environment B of the agent society. The input for the c-resistance is the same parameters as above.

22.5.3 Experiment B2

Here too, environment B is used. The input for the c-resistance is the same parameters as above with the addition of the monetary value of the current call.

22.6 Results

Table 22.1 Results for the best evolved functions in the three categories

Experiment	% lost calls	Collected revenue
A	6.8	404
B1	8.0	410
B2	7.6	411

As an example we give the function for experiment A. With distance d and load L, the c-resistance R_c is given by:

$$R_c = d + L \times (0.5 \times d + 2 + 0.5 \times d \times (d + L/(d - L^2)))$$

If the load is zero, the agents choose the shortest path, which, of course, is the correct solution. The expression between the parentheses cannot be explained, which is usually the case when applying evolutionary techniques.

The author hand-coded a number of c-resistance functions in order to find one that is superior to the best evolved function for environment A, but did not succeed. The best one, $d \times (1 + L^2)^2$, lost 8.0% of the calls.

It should be mentioned that agents, are insensitive to the loads, i.e. use the shortest-path algorithm, lose 11.2% of the calls in environment A.

Experiment B1 shows an especially interesting result: despite the agents' being blind to the fact that calls have different monetary values, they learn to improve their behaviour just through the feedback from their environment, the simulator. Though they lose more calls compared to experiment A, they lose relatively more of the cheaper ones, and so the overall performance is higher.

In experiment B2 the agents are given the ability to sense the different values and improve further, albeit only very little. We feel that the results, especially the result of experiment B1, are very promising.

22.7   Discussion and Future Work

Though comparisons with other solutions to the routing problem have not been conducted yet, the author firmly believes in the feasibility of the genetic programming approach. What we are studying here is an agent society in the sense that a large number of rather small and very similar software modules have to cooperate in order to fulfil a common goal in a certain environment. Thus, this research shows some features of swarm intelligence systems. These systems are mainly characterised by the fact that communication between the entities is indirect only, i.e. they do not send messages to each other but make marks on the environment that can be read by other entities.

In contrast, our agents also communicate directly, albeit only with a subset of the society members, i.e. the neighbours they are linked to. Moreover, at least at this stage of our work, communication is simplistic.

Our agents are mainly distinguished by their behaviour: given a set of sensors they make an assessment of the state of their environment by applying a function on the readings of their sensors. The outcome of the assessment determines their choice of actions.

Another feature is that they are able to learn, which is done off-line. As is the case with Artificial Neural Networks, they are trained in a training environment and are not able to change their basic behaviour when confronted with a different environment.

The next steps of our research take the following directions:

1. *Providing the agents with more sensors.* Primarily, the derivative of the load over time seems to be a reasonable candidate. Further, it is obvious that an agent's position in the network with regard to its topology should make a big difference to the behaviour. Nodes have between 3 and 5 neighbours, and their distance to the farthest node ranges from 2, for the central node, to 4, for some peripheral nodes. So, a node's arity and/or eccentricity will be candidate parameters for the c-resistance function.
2. *Specialisation of the agents.* An alternative to the introduction of topological parameters is to differentiate the agents, i.e. to evolve 13 different c-resistance functions, one for each agent, in the extreme case. A compromise would be to have the same c-resistance function for all nodes with the same topological properties. Even then the demands on computing capacity will be at least an order of magnitude higher compared to the topological parameter solution as described above.

3. *Changing the environment.* We will investigate how the performance of the agent society changes by:

- Changing the topology of the network by removing a link or node
- Changing the demand on it by feeding it with more or less different call patterns compared to the one used in the training environment.

Acknowledgements

The author gratefully acknowledges support from the Swedish Research Council for Engineering Sciences, NUTEK.

22.8   References

Appleby, S. and Steward, S. (1994) Mobile Software Agents for Control in Telecommunications Networks, *BT Journal of Technology*, **12(2)**, 104–113.

Arnold, W., Hellendoorn, H., Seising, R., Thomas, C. and Weitzel, R. (1997) *Fuzzy Routing: Application of Soft Computing Techniques to the Telecommunication Domain.* ERUDIT Service Center, Aachen, Germany.

Banzhaf, W., Nordin, P., Keller, R.E. and Francone, F.D. (1998) *Genetic Programming – An Introduction.* Morgan Kauffmann.

Bertsekas, D. and Gallager, R. (1992) *Data Networks.* Prentice-Hall.

Frei, C. and Faltings, B. (1998) A Dynamic Hierarchy of Intelligent Agents for Network Management, in *Proceedings of Intelligent Agents for Telecommunications Applications* (IATA'98), Springer-Verlag.

Gibney, M.A. and Jennings, N.R. (1998) Dynamic Resource Allocation by Market-Based Routing in Telecommunications Networks, in *Proceedings of Intelligent Agents for Telecommunications Applications* (IATA'98), pp. 102–117, Springer-Verlag.

Koza, J.R., Goldberg, D.E., Fogel, D.B. and Riolo, R.L. (Eds.) (1997) *Proceedings of the First Annual Conference on Genetic Programming*, MIT Press.

Lukschandl, E., Holmlund, M., Modén, E., Nordahl, M. and Nordin P. (1998a) Automatic Evolution of Java Bytecode with Genetic Programming: First Experience with the Java Virtual Machine, in *Late Breaking Papers at the First European Workshop on Genetic Programming* (EuroGP '98, Paris).

Lukschandl, E., Holmlund, M., Modén, E., Nordahl, M. and Nordin P. (1998b) Induction of Java Bytecode with Genetic Programming, in J.R. Koza (Ed.) *Late Breaking Papers at the Genetic Programming Conference,* Stanford University Bookstore.

Lukschandl, E., Borgvall, H., Nohle, L., Nordahl, M. and Nordin, P. (1999) Evolving Routing Algorithms with the JBGP-System, in *Proceedings of the First European Workshops, EvoIASP'99 and EuroEcTel'99*, Gothenburg, Sweden, Springer.

Schoonderwoerd, R, Holland, O.E. and Bruten, J. (1997) Ant-like Agents for Load Balancing in Telecommunications Networks, in *Proceedings of the 1st International Conference on Autonomous Agents*, ACM Press.

Steenstrup, M.E. (Ed.) (1995) *Routing in Communication Networks.* Prentice-Hall.

Index